Recent Progress on Numerical Analysis for Nonlinear Dispersive Equations

equations, highlighting the novel use of complex coefficients alongside traditional real coefficients, and provides detailed distinctions among various symmetric and conjugate schemes for long-term computations applied to some famous models. Chapter 4 presents a uniform-in-time first-order convergence result for the filtered Lie-Trotter time-splitting discretization of the mass-supercritical, energy-subcritical nonlinear Schrödinger equation, achieved through discrete Strichartz estimates and techniques adapted from nonlinear scattering theory, enabling the extension of local error estimates to a global uniform-in-time framework applicable for all (positive) time. Chapter 5 introduces new algorithms for the nonlinear cubic Schrödinger equation based on modulated solitons, revealing a completely integrable system for single soliton evolution and offering effective numerical approximations for multi-soliton interactions using the Dirac-Frenkel variational principle without requiring a complete discretization grid. Chapter 6 reviews novel exponential-type integrators utilizing frequency analysis for solving KdV-type equations with rough data, demonstrating that these methods achieve accurate solutions with lower regularity requirements compared to traditional approaches.

Rémi Carles and Chunmei Su

https://doi.org/10.1142/9789819816620_fmatter

Contents

https://doi.org/10.1142/9789819816620_0001

Chapter 1

A Unified Approach to Low-Regularity and Symmetric Time Integrators for Nonlinear Dispersive PDEs

Yvonne Alama Bronsard

Univ Rennes, IRMAR-UMR 6625, F-35000 Rennes, France
yvonne.alamabronsard@univ-rennes.fr

In recent years, there has been a growing interest in schemes constructed to reduce the smoothness assumptions on the design and analysis of numerical methods, in order to treat low-regularity dynamics. On the other hand, structure-preserving algorithms are essential in maintaining the intrinsic structure of the underlying equation. Combining low-regularity numerical approximations with structure-preserving algorithms offers a powerful framework for obtaining accurate simulations while ensuring that the fundamental physical properties of the system are conserved on the discrete level. The challenge lies in developing schemes that can effectively balance both low-regularity requirements and the preservation of structure, providing reliable approximations even in the presence of rough solutions. We give an overview of existing results addressing these topics, from the development of low-regularity schemes and their higher order extensions, to the design of novel symmetric schemes recently introduced in the literature.

Contents

1. Introduction

Algorithms preserving the structure of the underlying equation constitute an important part of computational mathematics, see [1, 2]. However, in general they heavily rely on highly regular solutions. For instance, consider the cubic nonlinear Schrödinger equation (NLS) equation:

$$\partial_t u(t,x) - i\Delta u(t,x) = -i|u(t,x)|^2 u(t,x). \tag{1.1}$$

Approximating the above in time using either finite difference schemes [3–5] which preserve the mass and energy, or Strang splitting methods [6, 7] which preserve the mass and nearly preserve the energy, require at least H^4 solutions for second-order convergence in L^2.

Recently, a novel class of *symmetric* schemes – built upon *resonance based* or *low-regularity* integrators – has been developed. These schemes not only preserve the underlying structure of the equation in numerical simulations but also achieve convergence for less regular solutions. In this chapter we start by presenting a toolbox allowing for the understanding of the nonlinear approaches and low-regularity approximation techniques used for constructing resonance based and low-regularity integrators existing in the literature. The tools thereby introduced serve as a foundation, allowing for the construction of these recently obtained symmetric schemes. We further show in numerical experiments the very good long-time behavior of the schemes, both in smooth and non-smooth settings. Surprisingly, even in the smooth setting, the schemes surpass the long time energy preservation properties of classical splitting methods. We begin by discussing the numerical treatment of low-regularity dynamics for dispersive equations.

The study of non-smooth phenomena, arising in nonlinear optics, shallow water waves, and wave turbulence, is at the heart of an intense international activity. While linear problems or smooth solutions are generally well understood, a reliable description of *nonlinear* and *non-smooth* phenomena remains an important challenge in computational mathematics today. This is especially the case when studying nonlinear dispersive equations, since one cannot expect the smoothing effects that are present in parabolic equations. The lack of smoothness in the solution leads to large errors in numerical simulations, massive computational costs and ultimately causes the failure of classical schemes. Yet many interesting physical phenomena exhibit non-smooth behavior. An essential task is thus to develop and study suitable numerical schemes for approximating such non-smooth behavior.

To this end, we will be interested in discussing existing works which develop numerical tools for approximating the following class of dispersive equations

$$\begin{cases} \partial_t u(t,x) - \mathcal{L}u(t,x) = p\big(u(t,x)\big), \ (t,x) \in \mathbb{R}\times\mathbb{T}^d, \ d \in \{1,2,3\}, \\ u(0,x) = u_0(x), \end{cases} \tag{1.2}$$

where $p\big(u(t,x)\big)$ is a polynomial in $u(t,x)$, its conjugate $\bar{u}(t,x)$, and their spatial derivatives. Throughout this introduction we ask $\mathcal{L}$ to be a linear operator which generates a strongly continuous group $\{e^{t\mathcal{L}}\}_{t\in\mathbb{R}}$ of contractions on L^2, and require $\mathcal{L}$ to involve only polynomials in the frequencies when mapped in Fourier space. Namely, we denote $P(k)$ the Fourier multiplier of $\mathcal{L}$ and require $P(k)$ to be polynomial in k, see also (5.5). We give two prototypical examples of nonlinear dispersive equations belonging to the above class (1.2).

Example 1. A classical model in quantum mechanics and a natural starting point for the study of dispersive equations is the NLS equation (1.1), which corresponds to taking $\mathcal{L} = i\Delta$, $P(k) = -i|k|^2$, and $p(u) = -i|u|^2u$.

Example 2. Another example is the Korteweg–de Vries equation (KdV), which serves as a model for waves in shallow waters, and corresponds to the case $\mathcal{L} = -\partial_x^3$, $P(k) = ik^3$, and $p(u) = \frac{1}{2}\partial_x\left(u^2\right)$:

$$\partial_t u(t,x) + \partial_x^3 u(t,x) = \frac{1}{2}\partial_x\left(u^2(t,x)\right). \tag{1.3}$$

We give an overview of recent techniques allowing for the development of computational schemes for approximating low regularity solutions to such equations, with an emphasis on what can be rigorously proven to guarantee their reliability. Numerically approximating low regularity solutions is a challenge since classical approximation techniques require smoothness. As an illustration, in the case of the NLS equation (1.1), both classical finite differences and splitting methods require at least H^2-solutions for L^2-convergence at the first order, and H^4-solutions for L^2-convergence at the second order [4–8]. On the other hand, the equation is locally well-posed (LWP) for solutions in $H^\alpha(\mathbb{T}^d)$, with $\alpha \geq 0$, $\alpha > 0$, and $\alpha > 1/2$ for $d = 1, 2$ and 3 respectively. A question of interest is how to reduce the gap between the degree of regularity required by the numerical approximation and the regularity needed for the local well-posedness theory.

The numerical challenges of discretizing such equations are in part governed by the theoretical challenges in proving their well-posedness.

Early well-posedness studies often focused on local well-posedness for relatively smooth initial data, where the analysis could often be reduced to linear problems. However, the complexity in the analysis increases significantly when considering solutions that exist over large times or when the initial data lack regularity. In such cases, understanding the intricate interactions of nonlinear waves becomes essential. The same can be said in the discrete setting. The goal of this chapter is to enhance our understanding of both aspects, by presenting the construction and analysis of integrators from the literature that demonstrate convergence with less regular data while maintaining favorable long-term behavior. Sections 2–5 focus on obtaining and analyzing schemes that converge under less regular data and Section 6 discusses their symmetric extensions and long-time behavior.

We begin with a comparison between *classical* approximation techniques and the recent *low-regularity* approaches.

2. A first-order scheme: Classical versus low-regularity approaches

A large toolbox of numerical schemes for dispersive equations has been developed, based on different discretization techniques. *Exponential integrators* and *Lawson methods* rely on discretizing Duhamel's formula, otherwise called the variation-of-constants formula. *Finite difference* schemes directly discretize the underlying equation (1.2) via explicit or implicit Taylor approximations. *Splitting methods* are based upon splitting the full equation into a series of simpler subproblems. The above classical schemes often allow for a precise and efficient approximation, but are designed for *smooth* solutions, requiring high regularity assumptions. To illustrate this, we will construct and compare first-order time-approximations to the class of dispersive equations (1.2). We start by writing *Duhamel's formula* for $u(t)$:

$$u(t) = e^{t\mathcal{L}}u_0 + \int_0^t e^{(t-s)\mathcal{L}}p\big(u(s)\big)ds, \tag{2.1}$$

and iterate Duhamel's formula for $u(s)$ inside $p\,(u(s))$ to obtain the next iterate

$$u(t) = e^{t\mathcal{L}}u_0 + \int_0^t e^{(t-s)\mathcal{L}}p(e^{s\mathcal{L}}u_0)ds + \int_0^t\int_0^s \dots ds_1 ds. \tag{2.2}$$

Namely, the above equality (2.2) follows by Taylor expanding the nonlinearity p appearing in (2.1) around the linear flow: $p\big(u(s)\big) = p(e^{s\mathcal{L}}u_0) + \int_0^s \dots ds_1$, with an integral remainder. We search an approximation over one time step $t = \tau$. In order to obtain a first-order approximation in *time*, we neglect the double integral in (2.2), which corresponds to the second-order term:

$$u(\tau) = e^{\tau\mathcal{L}}u_0 + \int_0^\tau e^{(\tau-s)\mathcal{L}}p(e^{s\mathcal{L}}u_0)ds + O(\tau^2). \tag{2.3}$$

If we suppose that $p(u)$ is locally Lipschitz, neglecting the $O(\tau^2)$-term does not require any additional regularity. As we are considering time approximations, the main challenge is to replace the integrand

$$e^{-s\mathcal{L}}p(e^{s\mathcal{L}}u_0) \tag{2.4}$$

by an approximation which can be integrated explicitly on $[0,\tau]$. Classical methods are based upon Taylor expanding in time the exponentials appearing in the above equation (2.4) as follows: $e^{s\mathcal{L}}u = 1 + O(s\mathcal{L}u)$. Hence, they are based upon linear frequency approximations. We give the example of a first-order exponential integrator.

Example 3 (The Euler exponential method). A first-order exponential integrator consists in Taylor expanding the exponential appearing inside the nonlinearity :

$$e^{-s\mathcal{L}}p(e^{s\mathcal{L}}u_0) \approx e^{-s\mathcal{L}}p(u_0). \tag{2.5}$$

The scheme after one time-step, which expresses u^{n+1} in terms of u^n, is then obtained by integrating exactly the remaining term $e^{(\tau-s)\mathcal{L}}$:

$$u^{n+1} = e^{\tau\mathcal{L}}u^n + \tau\varphi_1(\tau\mathcal{L})p(u^n) \quad \text{with} \quad \varphi_1(z) = \frac{e^z - 1}{z}. \tag{2.6}$$

It remains to quantify the local error induced by the approximation (2.5). For the NLS equation (1.1), following the reference [9], the local error has the following explicit form:

$$|u|^4u - 2|u|^2\Delta u + u^2\overline{\Delta u}. \tag{2.7}$$

First-order convergence of this method thus requires two additional derivatives on the solution, and more specifically requires H^2-solutions when measuring the error in L^2. If the solution is however not sufficiently smooth, the local error becomes unbounded and the exponential integrator fails to converge at the first order, see Figure 1.

A new discretization technique was introduced in the works [10, 11] in the case of the KdV equation (1.3) and the NLS equation (1.1). The idea is based on a new nonlinear approach, and a refined analysis made in Fourier space. Instead of linearizing the frequency interactions in the

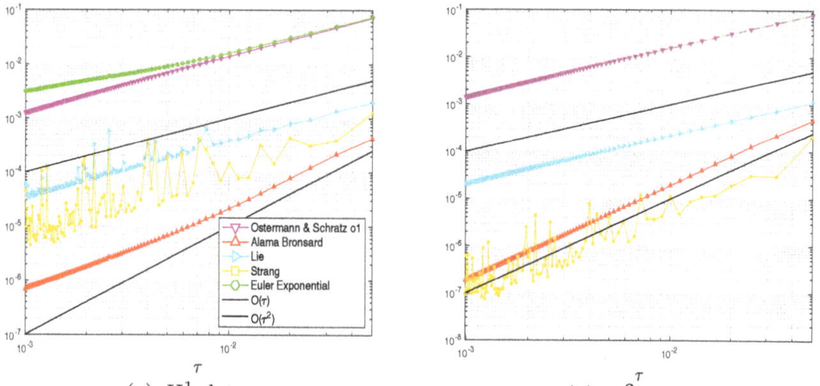

(a) H^1 data. (b) H^2 data.

Fig. 1. Convergence plots taken for classical integrators (green, blue, yellow) versus integrators suited for low-regularity solutions (pink and red). The Euler exponential method (2.6) in green, and the first-order Lie splitting (2.24) in blue do not converge at the first order in Plot 1(a). The second-order Strang splitting in yellow oscillates erratically in both plots, due to the lack of regularity. The first-order resonance-based scheme of [11] in pink (Ostermann & Schratz o1) is given by (2.20). The symmetric scheme of [12] in red (Alama Bronsard) is given by (6.2) and introduced in Section 6.1.

central oscillatory term (2.4), the nonlinear frequency interactions between the linear flow $e^{-s\mathcal{L}}$ and the nonlinear factor $p(e^{s\mathcal{L}}u_0)$ are incorporated into the numerical discretization. This allows for the convergence of novel time discretizations under lower regularity assumptions on the solution.

In the following section we illustrate the new discretization technique of [11] for the NLS equation (1.1), and compare it with classical approaches.

2.1. *Illustration on the nonlinear Schrödinger equation*

We propose in this section a method of construction introduced in [12] for the NLS equation (1.1), which offers a framework for the derivation of different classes of schemes: the previous scheme of [11] built at low-regularity, newly designed ones presented in this manuscript, and classical integrators. We recall that $p(u) = -i|u|^2 u$ and let $t_n = n\tau$, with τ the time step.

We start by making a change of variables. Instead of working with the equation (1.1) in u, we twist the variables with respect to the leading differential operator $\mathcal{L} = i\Delta$:

$$e^{it\Delta}v = u, \tag{2.8}$$

and work with the equation in v:

$$\partial_t v = e^{-it\Delta} p(e^{it\Delta} v), \quad v(0) = u_0, \tag{2.9}$$

as well as its integral form (obtained by integrating on $[t_n, t_n + s]$):

$$v(t_n + s) = v(t_n) + e^{-it_n\Delta} \int_0^s e^{-is_1\Delta} p(e^{i(t_n+s_1)\Delta} v(t_n + s_1)) ds_1. \tag{2.10}$$

This change of variables is a well-known approach in the theoretical analysis of dispersive equations in low regularity spaces [13], in rough path theory [14], as well as in numerical analysis for constructing Lawson methods [15, 16] and resonance-based schemes [10, 11].

Remark 4 (The advantage of twisting the variable). The equation (1.1) for u is driven by the differential operator $i\Delta$, an unbounded operator on L^2. This is not the case for the equation (2.9) in v, which is written in terms of the nonlinearity and the free Schrödinger group, a bounded operator in L^2. The numerical advantage thus lies in the fact that the approximation

$$v(t_n + s) \approx v(t_n), \quad |s| \leq \tau, \tag{2.11}$$

does not require additional regularity on the solution v. Indeed, it follows from equation (2.10) that for $\sigma > d/2$,

$$\begin{aligned} \|v(t_n + s) - v(t_n)\|_{H^\sigma} &\leq \int_0^s \|p(e^{i(t_n+s_1)\Delta} v(t_n + s_1))\|_{H^\sigma} ds_1 \\ &\leq \left(\sup_{0 \leq s_1 \leq s} \|v(t_n + s_1)\|^3_{H^\sigma} \right) s. \end{aligned}$$

The same analysis would yield in the untwisted variable the approximation $u(t_n + s) \approx e^{is\Delta} u(t_n)$, $|s| \leq \tau$. Obtaining the approximation $u(t_n + s) \approx u(t_n)$ requires Taylor expanding at the first order the linear flow $e^{is\Delta}$, and hence two additional derivatives on the solution.

Next, we map Duhamel's formula (2.10) in Fourier space, with $s = \tau$ the time step. Expanding v in Fourier series $v(t, x) = \sum_{k\in\mathbb{Z}} v_k(t) e^{ikx}$, allows to express the action of the linear Schrödinger flow on v, $e^{\pm it\Delta} v(t, x) = \sum_{k\in\mathbb{Z}} v_k(t) e^{\mp itk^2} e^{ikx}$. We note that the upcoming derivation can analogously be obtained in higher dimensions with $\mathbf{k} = (k^1, \cdots, k^d) \in \mathbb{Z}^d$. The k-th Fourier coefficient of $v(t_{n+1})$ reads

$$\begin{aligned} v_k(t_{n+1}) = v_k(t_n) - i &\sum_{k=-k_1+k_2+k_3} e^{it_n R(k)} \\ &\times \int_0^\tau e^{isR(k)} \overline{v_{k_1}(t_n + s)} v_{k_2}(t_n + s) v_{k_3}(t_n + s) ds, \end{aligned} \tag{2.12}$$

where we collected all the frequency interactions into the *resonance structure* $R(k)$:

$$\begin{aligned} R(k) &= k^2 + k_1^2 - k_2^2 - k_3^2 = 2k_1^2 - 2k_1(k_2+k_3) + 2k_2k_3, \\ k &= -k_1 + k_2 + k_3. \end{aligned} \tag{2.13}$$

The goal is to approximate in time the integrand in equation (2.12). The idea behind the novel nonlinear approach in [11], is to base the approximation on the specific form of the resonance structure (2.13). We illustrate this approach with the general decomposition presented in [12], which improves on the scheme of [11]. Namely, we decompose the integrand in (2.12) into three factors:

$$I_\tau = \int_0^\tau e^{isR_{\mathrm{dom}}(k)} e^{isR_{\mathrm{low}}(k)} g(s)ds, \tag{2.14}$$

with

$$R(k) = R_{\mathrm{dom}}(k) + R_{\mathrm{low}}(k), \quad g(s) = \overline{v}_{k_1}(t_n+s) v_{k_2}(t_n+s) v_{k_3}(t_n+s), \tag{2.15}$$

$$R_{\mathrm{dom}} = 2k_1^2, \quad R_{\mathrm{low}} = -2k_1(k_2+k_3) + 2k_2k_3. \tag{2.16}$$

The choice of the above decomposition (2.14) is dictated by the regularity assumptions needed to approximate each of the factors. Using Duhamel's formula (2.10) we can approximate $g(s)$ at its left end point: $g(s) \approx g(0)$ with a local error in $O(s)$ *without additional regularity requirements*, see equation (2.11). It remains to treat the oscillatory term:

$$e^{isR_{\mathrm{dom}}(k)} e^{isR_{\mathrm{low}}(k)} g(0), \tag{2.17}$$

which corresponds to the term (2.4), when mapped in Fourier space. A second approximation step is necessary to obtain a practical implementation of the scheme, see Remark 5. We observe that approximating the factor $e^{isR_{\mathrm{low}}(k)}$ asks for *less* regularity than approximating $e^{isR_{\mathrm{dom}}(k)}$. Indeed, the terms $k_l k_j$ (for $l \neq j$) appearing in $R_{\mathrm{low}}(k)$ correspond to products of first-order derivatives. Hence, by denoting $v = v(t_n)$, a formal first-order Taylor expansion yields

$$e^{isR_{\mathrm{low}}} = 1 + O(sR_{\mathrm{low}}), \quad \text{with} \quad R_{\mathrm{low}} g(0) \sim -2(|\nabla v|^2 \overline{v} + 2v \nabla v \cdot \nabla \overline{v}), \tag{2.18}$$

after mapping back from frequency to spatial variables. If we tried to approximate the dominant factor $e^{isR_{\mathrm{dom}}(k)}$, as $2k_1^2$ corresponds to second-order derivatives in Fourier space, we would require more regularity:

$$e^{isR_{\mathrm{dom}}} = 1 + O(sR_{\mathrm{dom}}), \quad \text{with} \quad R_{\mathrm{dom}} g(0) \sim 2v^2 \Delta \overline{v}. \tag{2.19}$$

Therefore, we choose to Taylor expand the lower term (2.18), making the approximation $R(k) \approx R_{\text{dom}}(k)$, and *integrate exactly* the dominant term:

$$I_\tau \approx \int_0^\tau e^{isR_{\text{dom}}(k)} ds\, g(0) = \frac{e^{2i\tau k_1^2} - 1}{2ik_1^2} g(0),$$

in order to obtain a scheme whose local error requires the least regularity. We finally map back in physical space, and in the untwisted variables. We thereby obtain the so-called first-order *resonance-based* integrator

$$u^{n+1} = e^{i\tau\Delta} u^n - i\tau e^{i\tau\Delta} \left((u^n)^2 \varphi_1(-2i\tau\Delta)\overline{u}^n \right), \quad \varphi_1(z) = \frac{e^z - 1}{z}. \tag{2.20}$$

The local error of the scheme is governed by the approximation made in (2.18) and is of the form:

$$-2(|\nabla v|^2 \bar{v} + 2v\nabla v \cdot \nabla \bar{v}), \tag{2.21}$$

which only requires the boundedness of one, instead of two, additional derivative on the solution. For a discussion on the practical implementation of the above scheme, we again refer to Remark 5, and for the terminology *resonance-based* integrator to Remark 6.

Remark 5 (An interplay between efficient implementation and proximity to the solution). Ideally, once the first approximation step $g(s) \approx g(0)$ is performed, one would hope to integrate the remaining oscillations (2.17) exactly in time (on $[0, \tau]$). While this can be done analytically:

$$\int_0^\tau e^{iR(k)s} ds = \frac{e^{i\tau R(k)} - 1}{iR(k)}, \quad R(k) = 2k_1^2 - 2k_1(k_2 + k_3) + 2k_2k_3, \tag{2.22}$$

it does not yield a practical implementation, as it results in a generalized convolution (of Coifman-Meyer type [17]) which cannot be mapped back to physical space, due to the factor $\left(2k_1^2 - 2k_1(k_2 + k_3) + 2k_2k_3\right)^{-1} \bar{v}_{k_1} v_{k_2} v_{k_3}$. Hence, the computations would need to be fully made in Fourier space. Carrying this out in higher spatial dimensions d would cause large memory and computational efforts of order $O(|K|^{2d})$, where K denotes the highest frequency in the discretization. For practical computations, we want to be able to express the discretization also in physical space in order to use the Fast Fourier Transform (FFT) which is of computational effort of order $O(|K|^d log|K|^d)$. Therefore, we choose an approximation of the integrand (2.17) which allows for a practical implementation (by not performing exact integration), while optimizing the local error to require the least regularity.

The latter dictates the choice of the decomposition (2.15). If one was not concerned with the lack of regularity of the solution, but for example the conservation of a certain quantity of interest, then the choice of the frequency decomposition (2.15) could be taken differently.

Remark 6 (The scheme entitled "resonance-based integrator"). The terminology *resonance-based* integrator comes from the fact that the scheme is based upon the resonance structure (2.13) of the underlying PDE. Another name used is *Fourier* integrator. Further on, we will discuss extensions to these results which are suited on more general domains. These integrators will then be termed *low regularity* integrators.

We would like to address a common misconception. The construction of these resonance-based integrators are not based on the resonant case $R(k) = 0$. From a numerical point of view, the resonant case $R(k) = 0$ is trivial given that there are no oscillations to approximate; the discretization of Duhamel's integral would simply amount to $\tau p(u^n)$. However, the case $R(k) \neq 0$ is more subtle as one needs to capture the oscillations $e^{isR(k)}$ also in the numerical approximation. We note that interestingly, the difficulties are reversed when working in the theoretical analysis of dispersive equations, since in the resonant case $R(k) = 0$ we do not gain the decay $\frac{1}{R(k)}$ in the frequencies when integrating (see equation (2.22)).

The above scheme (2.20) was first obtained by [11], using slightly different steps. The authors of [11] do not pass directly in Fourier space – as in (2.12) – and first approximate $v(t_n + s_1) \approx v(t_n)$ in Duhamel's integral (2.10), as this does not require additional regularity (see (2.11)). However, making this first approximation ($v(t_n + s) \approx v(t_n)$ or in Fourier space $g(s) \approx g(0)$) breaks the symmetric structure of the underlying PDE (1.1) and leads to loss of structure preservation. We will see in Section 6 that to obtain schemes which have good structure-preserving properties we need to make a different approximation to $v(t_n + s)$ (or in Fourier $g(s)$) and should consider the general form (2.14). This approach was first taken in [12].

We refer to Figure 1(a) where first-order convergence of the resonance-based scheme (2.20) in L^2 norm for H^1 solutions is observed in numerical experiments. In Section 3.1 rigorous convergence results of the scheme are discussed.

We next compare this resonance-based discretization with classical time integrators.

Comparison with classical approximation techniques. The Euler exponential method corresponds to making the approximation $R(k) \approx k^2$ recovering the scheme (2.6) with local error (2.7).

Another quite popular method of approximation is splitting method [2, 6, 7, 18]. We recall that splitting method consists in separating the linear $\mathcal{L} = i\Delta$ and nonlinear part $p(u) = -i|u|^2u$ and composing the exact flows of the differential equations:

$$\partial_t u = i\Delta u, \quad \text{and} \quad \partial_t u = -i|u|^2 u. \tag{2.23}$$

The first-order Lie splitting scheme is then given by:

$$\begin{aligned} u^{n+1/2} &= e^{-i\tau|u^n|^2} u^n, \\ u^{n+1} &= e^{i\tau\Delta} u^{n+1/2}, \end{aligned} \tag{2.24}$$

with the explicit local error

$$2\left(2|\nabla u|^2 u + \bar{u}(\nabla u \cdot \nabla u) + u^2 \Delta \bar{u}\right), \tag{2.25}$$

see for example [6, Section 4.2]. Because of the last term $u^2\Delta\bar{u}$ in the above formula, the Lie splitting method requires two additional derivatives on the solution, as does the Euler exponential method.

To understand the difference behind this splitting approximation (2.24) and the resonance-based approach (2.14), we formally Taylor expand the nonlinear part in (2.24): $e^{-i\tau|u|^2} = 1 - i\tau|u|^2 + O(\tau^2)$. This does not require additional regularity assumptions since the error term in $O(\tau^2)$ only involves a polynomial of u. The local error analysis is then dominated by the remainder

$$u(t_{n+1}) - \left(e^{i\tau\Delta}u(t_n) - i\tau e^{i\tau\Delta}\left(|u(t_n)|^2 u(t_n)\right)\right). \tag{2.26}$$

In the frequency variables, the above corresponds to taking $R(k) \approx 0$ (and $g(s) \approx g(0)$), thereby neglecting all the oscillations in (2.17). Equivalently, by recalling Duhamel's integral (2.3) written in the spatial variables, the splitting method corresponds to the approximation

$$e^{-s\mathcal{L}}p(e^{s\mathcal{L}}u_0) \approx p(u_0), \tag{2.27}$$

which treats the problem as a linear one. Such linearized frequency approximations are very efficient to implement, since the oscillations in (2.27) no longer appear in the approximation. However, the method does not resolve the nonlinear frequency interactions of the system and treats the resolution of the linear $i\Delta$ and the nonlinear term $p(u)$ independently.

In conclusion, these classical approximation techniques do not take into account the interactions between the frequencies from the linear and nonlinear terms. The idea behind the resonance-based schemes is to do so, while maintaining a scheme that can be efficiently implemented at the same computational cost as classical schemes. In the course of this manuscript, we will build on this nonlinear approach to obtain both novel low-regularity approximations and better structure preserving schemes.

The next section is devoted to a discussion on prior rigorous convergence results and on the spaces in which they are obtained, both for time integrators as well as finite difference methods in time.

3. Explicit convergence rates

In this section we discuss how to obtain explicit convergence rates of order τ^ν, which are quantified as a function of the Sobolev regularity of the solution, $u \in C([0,T], H^\alpha)$, and the Sobolev norm H^σ in which we measure the error. We let $\nu > 0$ and $\alpha \geq \sigma \geq 0$ and consider the error:

$$\|u(t_n) - u^n\|_{H^\sigma} \leq C_T \left(\sup_{[0,T]} \|u(t)\|_{H^\alpha} \right) \tau^\nu, \quad 0 \leq t_n = n\tau \leq T. \tag{3.1}$$

We denote $C_T(\cdot)$ by a generic constant which depends continuously on its arguments, and exponentially on the final time T. These error analysis results are thereby *local in time*, see Remark 15. Unless stated otherwise we set ourselves on $\mathbb{T}^d$, with $d \in \{1,2,3\}$. We present two different perspectives when exploring the interplay between ν, α and σ:

(1) *Full order of convergence.* We fix the desired order of convergence $\nu = m \in \mathbb{N}$, and search for the optimal regularity $\alpha = \alpha(d,\sigma,\nu)$ which is needed for the convergence of the method to order ν. The required regularity $\alpha(d,\sigma,\nu)$ depends on the spatial dimension d, the norm H^σ, and the order of convergence ν. This approach is the most common one in the literature.

(2) *Fractional order of convergence.* We fix the H^σ-norm, and assume the regularity $u \in C([0,T], H^\alpha)$. One cannot hope to obtain full order of convergence when the boundedness of the local error term requires higher regularity assumptions than what is assumed on the solution $u(t)$. Fractional convergence rates are then expected, with the rate as a function of the regularity α. This approach is more general, since we recover the full-order convergence result, when enough regularity is assumed.

The choice of the regularity parameters α and σ strongly influences the complexity of the analysis to obtain a convergence result of type (3.1). We distinguish three categories, from easiest to most challenging:

- *Case* $\sigma > d/2$. We need $\alpha \geq \sigma$ and hence also $\alpha > d/2$. In this case we have that $H^\sigma \hookrightarrow L^\infty$, and by using Young's convolution inequality together with Holder's inequality in Fourier space it follows that H^σ forms an algebra: $\|uv\|_{H^\sigma} \leq C_\sigma \|u\|_{H^\sigma} \|v\|_{H^\sigma}$. Both the local error and stability analysis follow easily for a polynomial nonlinearity, and applying Lady Windermere's fan argument allows to conclude global convergence.
- *Case* $0 \leq \sigma \leq d/2$ *and* $\alpha > d/2$. We have $u(t) \in L^\infty$, however the space H^σ does not form an algebra, and hence one needs refined bilinear estimates. In particular, for the stability of the scheme the following estimate is crucial for bounding the nonlinear terms:

$$||wz||_{H^\sigma} \leq C_{\sigma,d}||w||_{\frac{d}{2}+\epsilon}||z||_{H^\sigma}, \quad 0 \leq \sigma \leq d/2, \quad \epsilon > 0. \tag{3.2}$$

  Using a refined global error analysis, one can still push the error analysis down to H^σ through a Lady-Windermere's fan argument, by first proving fractional convergence of the scheme in a suitable higher order Sobolev space, see [6].
- *Case* $0 \leq \sigma \leq d/2$ *and* $0 < \alpha \leq d/2$. Given that an estimate of the form (3.2) is out of reach in this rough setting, we cannot obtain the global convergence by combining a local error and stability analysis through Lady Windermere's fan argument. As in the theoretical analysis of PDEs at low regularity, in order to consider very rough solutions $u(t) \in H^\alpha$ with $\alpha \leq d/2$, one needs to call upon more refined tools such as Strichartz estimates and Bourgain spaces [13, 19, 20].

The above cases are particularly relevant when the nonlinearity is a polynomial in u and $\bar{u}$, such as for the NLS equation (1.1), which is discussed next.

In the following Section 3.1, we discuss the state of the art regarding convergence results for the NLS equation, before branching out in more recent research directions in Section 4, which are detailed in the rest of this chapter.

3.1. *Comparing convergence results for the nonlinear Schrödinger equation*

Resonance-based schemes. We expect resonance-based schemes to have the following convergence rates:

Convergence in H^σ at order τ^ν for $u_0 \in H^\alpha$, with $\alpha = \sigma + \nu$ and $\sigma > d/2$.

We refer to [11] in the case $0 < \nu \leq 1$, and [21] in the case $\nu = 2$. In these results the analysis is restrained to the torus using Fourier series and the frequency decomposition (2.17).

In the case $0 \leq \sigma \leq d/2$ more regularity assumptions are needed and we refer to the work [22], which treats the cases $\sigma \geq 0$, $\nu = 1, 2$ and $\alpha > d/2$.

Finally, the authors of [23] introduce discrete Bourgain spaces allowing them to consider the rough case of $0 < \alpha \leq 1/2$ on the one-dimensional torus $\mathbb{T}$. They show fractional convergence in L^2 of order τ^ν for $u_0 \in H^\alpha(\mathbb{T})$ with $0 < \alpha \leq 1$ and $\nu < 1$.

When set on the one-dimensional torus $\mathbb{T}$, several recent results build on the approach of making a decomposition in Fourier space and use the resonance structure (2.13), either for coupling the time integrator with a spectral spatial discretization, or for designing schemes under even lower regularity assumptions, see [24–27].

Splitting and exponential integrators. In general, the order of convergence of these integrators behaves as follows:

Convergence in H^σ of order τ^ν for $u_0 \in H^\alpha$, with $\alpha = \sigma + 2\nu$ and $\sigma \geq 0$.

In the case $u_0 \in H^\alpha$ with $\alpha > d/2$, we refer to the works of [2, 6, 7, 28–30] for convergence of Lie and Strang splitting methods, and to [9, 16, 31] for exponential integrators. In the more subtle case $\alpha \leq d/2$, we refer to [32] and [33], which show fractional rates up to first order with $\sigma = 0$, $\nu = \frac{\alpha}{2}$, and $0 < \alpha \leq 2$, of a filtered version of the Lie splitting method.

Finite difference approximations in time. For an overview on these methods (with a comparison to splitting techniques) we refer to [34]. The convergence of these methods requires at least as much regularity as splitting or exponential methods. For example, second-order convergence of a leap-frog and relaxation type finite difference schemes are obtained in [5] and [4], respectively. They require at least four additional derivatives on the solution, as is the case of exponential and splitting methods.

Table 1. Convergence results of the form (3.1) for the NLS equation (1.1).

Type	Ref.	Domain	Order ν	Norm σ	$\frac{\alpha-\sigma}{\nu}$
Resonance-based & low-regularity schemes	[11]	$\mathbb{T}^d$	$(0, 1]$	$> d/2$	1
	[21]	$\mathbb{T}^d$	2	$> d/2$	1
	[22]	$\Omega \subset \mathbb{R}^d$	1, 2	$> d/2$	1
	[22]	$\Omega \subset \mathbb{R}^d$	1, 2	$[0, d/2)$	$1 + \frac{d/2-\sigma}{2\nu}$
Splitting & Exponential Integrators	[7]	$\mathbb{R}^2$	1, 2	0	2
	[6]	$\mathbb{R}^3$	2	0	2
	[29]	$\mathbb{R}^d$	$\mathbb{N}$	0	2
Finite differences	[4]	$\mathbb{R}^d$	2	$> d/2$	2

Note: The interplay between the error norm σ, the convergence rate γ, and the regularity assumption α is expressed by the value of $\frac{\alpha-\sigma}{\nu}$, which represents the additional regularity requirement per order of convergence.

4. Expanding on the resonance-based scheme for NLS

Following this introduction to resonance-based integrators, many questions remain.

(1) **General domains.** The construction of these resonance-based schemes entirely relies on decomposing the solution on its Fourier basis, and thereby on imposing periodic boundary conditions. A first natural question is whether one can generalize the construction of these schemes from the torus $\mathbb{T}^d$ to more general domains, such as smooth bounded domains $\Omega \subset \mathbb{R}^d$ with homogeneous Dirichlet boundary conditions.

(2) **Higher order.** Up to now we have only constructed the first-order resonance-based scheme (2.20), and discussed results of order at most two. A next question is whether we can push the design up to (arbitrarily) high order.

(3) **Unifying approach.** These resonance-based schemes are built on the resonance structure of the underlying equation, see (2.13) for NLS equation. Can one build an algorithmic procedure to construct these schemes for a class of equations, such as for the class (1.2), without the need of analyzing one equation at a time?

(4) **Structure preservation.** Finally, we are interested in obtaining schemes which exhibit good long-time behavior, preserving – up to a certain extent – the conserved quantities of the equation on the discrete level. The resonance-based scheme (2.20) unfortunately

does not have good *structure-preserving* properties. Is it possible to construct schemes at low-regularity which also preserves the geometric structure of the equation?

In Section 5 we discuss the work [35] which answers questions regarding topics 1, 2 and 3, by offering an algorithmic way to construct higher-order schemes to a class of nonlinear evolution equations (with potentials) set on general domains. This latter work generalizes the construction of the schemes made in the article [22] also presented in this section, which designs and analyses schemes for the Gross-Pitaevskii equation in low-regularity regimes.

Section 6 focuses on topic 4 and introduces schemes which are advantageous both in non-smooth *and* smooth settings. We first introduce the work [12] which presents a novel time integrator for the NLS equation (1.1). The integrator is *symmetric*, provably converges at low-regularity, and exhibits in numerical experiments remarkable long-time near-preservation of the mass and energy, without the need of any CFL type conditions. This section also addresses topic 1. We then present the work [36] which combines the topics 2, 3, and 4, by generalizing the latter work to the class of dispersive equations (1.2) up to high orders. In numerical simulations, we witness better structure preserving properties of these new symmetric integrators compared to classical splitting schemes, even in the smooth setting of C^∞-solutions.

We begin with the state of the art regarding topic 1.

5. Beyond periodic boundary conditions

Our starting point is the work [37], which overcomes the necessity of Fourier-based expansions, and hence of periodic boundary conditions, by introducing new filtering techniques and a functional setting based on semigroup theory. The authors introduce a framework allowing for the construction and analysis of first- and second-order *low-regularity integrators* on more general domains and to the following class of nonlinear evolution equations:

$$\begin{cases} \partial_t u - \mathcal{L}u = f(u, \overline{u}), & (t, x) \in \mathbb{R} \times \Omega, \\ u(0, x) = u_0(x), & x \in \Omega, \end{cases} \tag{5.1}$$

with Ω either equal to the full space $\mathbb{R}^d$, the torus $\mathbb{T}^d$, or a smooth bounded domain $\Omega \subset \mathbb{R}^d$. In the case of a smooth bounded domain, the problem is equipped with homogeneous boundary conditions, which are encoded in the choice of the domain of the operator $\mathcal{L}$. We let the operator $\overline{\mathcal{L}}$ be defined by $\overline{\mathcal{L}\bar{u}}$. The main assumptions on $\mathcal{L}$ are the following:

(i) $\mathcal{L}$ is a linear operator defined on a Hilbert space X of complex valued functions $u \in \mathbb{C}$;
(ii) $\mathcal{L}$ generates a strongly continuous semigroup $\{e^{t\mathcal{L}}\}_{t\geq 0}$ of contractions on X;
(iii) $\mathcal{A} = -\mathcal{L} + \overline{\mathcal{L}}$ generates a group $\{e^{t\mathcal{A}}\}_{t\in\mathbb{R}}$ of unitary operators on X;
(iv) $\mathcal{L}$ and $\overline{\mathcal{L}}$ commute: $[\mathcal{L}, \overline{\mathcal{L}}] = 0$.

Example 7. Admissible operators $\mathcal{L}$ include the heat operator Δ, the Schrödinger operator $i\Delta$, the half wave operator $i|\nabla|$, and the Klein-Gordon and wave type operators $i\sqrt{-\Delta + m^2}$, with mass $m \geq 0$.

In addition, it is assumed that the nonlinearity f is tensorized under the form

$$f(u, \bar{u}) = \mathcal{B}\left(F(u) \cdot G(\bar{u})\right), \quad F, G : \mathbb{C} \to \mathbb{C}^J, \tag{5.2}$$

where $\mathcal{B}$ is a linear operator. It is also assumed that the mapping $u \mapsto f(u, \bar{u})$ is locally Lipschitz and maps a ball into itself.

The class of evolution equations (5.1) encompasses the class of dispersive equations (1.2). Indeed, with the above framework, the authors of [37] can deal in a unified way with parabolic, hyperbolic, dispersive as well as mixed type equations. However, an important class of equations is missing, namely equations with a *potential*. The goal of the work [22], is to close this gap and propose a novel low-regularity integrator for the Gross-Pitaevskii (GP) equation,

$$i\partial_t u(t,x) = -\Delta u(t,x) + V(x)u(t,x) + |u(t,x)|^2 u(t,x), \quad (t,x) \in \mathbb{R} \times \Omega, \tag{5.3}$$

with a non-smooth potential $V(x)$. In this work first- and second-order integrators are constructed and are proven to converge at an integer order $\nu = m \in \{1, 2\}$, under low-regularity assumptions both on the solution $u(t)$

and on the potential V. We denote by $u^n = \Phi_m^\tau(u^{n-1})$ the low-regularity integrators of order m. Set on the d-dimensional torus $\mathbb{T}^d$, the convergence result of [22] reads as follows:

Theorem 8. *Let $d \in \{1,2,3\}$, $\sigma \geq 0$, $T > 0$, $m \in \{1,2\}$ and*

$$\alpha := \begin{cases} m + \sigma, & \text{if } \sigma > \frac{d}{2}, \\ m + \frac{\sigma}{2} + \frac{d}{4}, & \text{if } 0 \leq \sigma < \frac{d}{2}. \end{cases}$$

For every $u_0 \in H^\alpha$ and $V \in H^\alpha$, there exists $\tau_{\min} > 0$ and $C_T > 0$ such that for every $\tau \leq \tau_{\min}$:

$$\boxed{\|u(t_n) - u^n\|_{H^\sigma} \leq C_T \tau^m, \qquad 0 \leq t_n = n\tau \leq T.}$$

The admissible values of α and σ are given in the blue zone on the right picture. For simplicity, the error analysis was made on the torus $\mathbb{T}^d$. Nevertheless, the tools involved are inspired by the filtering techniques of [37], which do not use Fourier-based expansions and hence can be extended to more general domains.

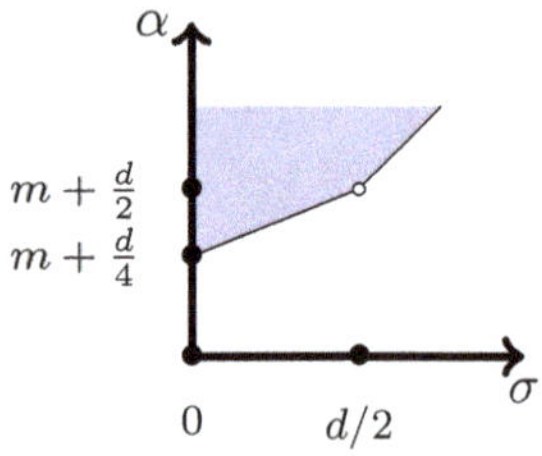

We refer to the work [12], which introduces a functional framework yielding convergence result both on the torus and on smooth bounded domains with homogeneous Dirichlet boundary conditions.

The work [22] motivated the work [35] presented in Section 5.1.2, which generalizes the construction of low-regularity schemes *up to arbitrary order* and for a class of nonlinear evolution equations with potentials.

In the next section we discuss topics 2 and 3.

5.1. *Higher-order approximations*

In order to obtain more precise approximations, which are also computationally efficient, one needs to construct higher-order methods. Developing such methods requires involved nontrivial calculations, especially when made at low regularity or in high dimensions, and is very laborious when the schemes are constructed for one equation at a time.

Based on iterating Duhamel's formula (2.1) once, we constructed in Section 2 the *first-order* resonance-based integrator (2.20) as well as the Euler exponential method (2.6). In order to obtain higher-order schemes one needs to include the approximation to higher-order Duhamel iterates:

$$u(t) = e^{t\mathcal{L}}u_0 + \int_0^t e^{(t-s_1)\mathcal{L}}p(e^{s_1\mathcal{L}}u_0)ds_1$$
$$+ \int_0^t e^{(t-s_1)\mathcal{L}}\left[p'(e^{s_1\mathcal{L}}u_0)\int_0^{s_1} e^{(s_1-s_2)\mathcal{L}}p(e^{s_2\mathcal{L}}u_0)ds_2\right]ds_1 + \ldots. \tag{5.4}$$

In the above we approximated p at $u(s_j)$ by its Taylor expansion about the linear flow $e^{s_j\mathcal{L}}u_0$, $j \in \mathbb{N}$:

$$p(u(s_j)) = p(e^{s_j\mathcal{L}}u_0) + p'(e^{s_j\mathcal{L}}u_0)\int_0^{s_j} e^{(s_j-s_{j+1})\mathcal{L}}p(e^{s_{j+1}\mathcal{L}}u_0)ds_{j+1} + \ldots.$$

We note that Taylor expanding $p(u(s_j))$ in the time variable was avoided, as this would require additional regularity on the solution.

We let $r+1$ be the desired order of the scheme. To obtain a scheme of order $r+1$, we need to consider the terms in the above formula (5.4) that contain up to $r+1$ integrals in time, and approximate them suitably.

We start by discussing higher-order approximations to the class of dispersive equations (1.2), where we assume that the differential operator $\mathcal{L}$ is real and that the differential operator $\mathcal{L}(\nabla)$ casts in Fourier space into the form

$$\mathcal{L}(\nabla)(k) = k^\sigma + \sum_{\gamma:|\gamma|<\sigma} a_\gamma \prod_j k_j^{\gamma_j}, \tag{5.5}$$

for some $\sigma \in \mathbb{N}$, $\gamma \in \mathbb{Z}^d$ and $|\gamma| = \sum_i \gamma_i$, where for $k = (k_1, \ldots, k_d) \in \mathbb{Z}^d$ and $m = (m_1, \ldots, m_d) \in \mathcal{Z}^d$ we set

$$k^\sigma = k_1^\sigma + \ldots + k_d^\sigma, \qquad k \cdot m = k_1 m_1 + \ldots + k_d m_d.$$

For an approximation at low-regularity using resonance-based schemes, the idea is to map the Duhamel iterates into Fourier space. The k-th Fourier coefficient of $u(t)$ is denote by:

$$u_k(t) = e^{tP(k)}u_k(0) + e^{tP(k)}\int_0^t e^{-s_1P(k)}p_k(e^{s_1P(k)}u_k(0))ds_1$$
$$+ e^{tP(k)}\int_0^t e^{-s_1P(k)}\left[p_k'(e^{s_1P(k)}u_k(0))e^{s_1P(k)}\right.$$
$$\left.\times \int_0^{s_1} e^{-s_2P(k)}p_k(e^{s_2P(k)}u_k(0))ds_2\right]ds_1 + \ldots, \tag{5.6}$$

where the Fourier multiplier of $\mathcal{L}$ is denoted by $P(k)$. For instance, for the Schrödinger equation we have $\mathcal{L} = i\Delta$ and $P(k) = -ik^2$. With a slight abuse of notation, we denoted by k all of the frequencies in Fourier, instead of k_1, k_2, k_3, etc.

At higher order, the resonance structure of the iterated integrals becomes very intricate, and it is a complicated task to choose a suitable resonance-based discretization. In the following, we discuss new techniques allowing for higher-order constructions.

5.1.1. *Decorated trees*

We first present the work [38], which introduces an algorithmic approach to build resonance-based schemes up to arbitrary order and to the class of dispersive equations (1.2) on the d-dimensional torus. Generally speaking, the work [38] develops an algebraic framework – inspired by those in SPDEs via regularity structures [39, 40] – to systematize a new analytical idea. Namely, they encode the scheme using combinatorial structure called *decorated trees*, allowing to systematize the approximation of the Duhamel iterates. We divide the idea behind the work [38] into two steps:

[a.] *On the continuous level.* We encode each of the Duhamel iterates in (5.6) by a decorated tree. In the case of the NLS equation (1.1), we have:

$$u_k(t) = \underbrace{e^{tP(k)}}\, u_k(0) + e^{tP(k)} \underbrace{\int_0^t e^{-s_1P(k)} p_k(e^{s_1P(k)}u_k(0))ds_1} + e^{tP(k)}$$

Π  (t) $\qquad$ Π  (t)

$$\underbrace{\int_0^t e^{-s_1P(k)}\Big[p_k'(e^{s_1P(k)}u_k(0))e^{s_1P(k)}\int_0^{s_1} e^{-s_2P(k)} p_k(e^{s_2P(k)}u_k(0))ds_2\Big]ds_1}$$

Π 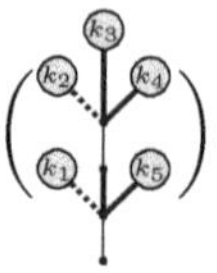 (t)

$$+\ldots, \tag{5.7}$$

where Π maps *decorated trees* to *Duhamel iterates.* The edges encode the operators in Duhamel's formula. Namely, the brown edge encodes the linear flow $e^{tP(k)}$, and the blue edge encodes the time integral $\int_0^t e^{-s_1P(k)}\ldots\, ds_1$.

A dotted edge corresponds to taking the complex conjugate of the expressions. For example, we have

$$\Pi(\text{}$$$$)(t) = e^{tP(k)}, \quad \Pi(\text{}$$$$)(t) = e^{-tP(k)}.$$

Lastly, the node decoration encodes the frequencies. This approach of encoding Duhamel's iterates in Fourier using decorated trees is also widely used in the theoretical study of dispersive equations.

The character Π, called a *pre-model* or *Feynman rule*, separates the algebraic level from the analytical one, and allows to expand the solution as a tree series expansion. We let $r+1$ denote the order of the scheme, and $\mathcal{V}_k^r$ the set of trees of size $r+1$, encoding an iterated integral with up to $r+1$ integrals. We can expand the solution (truncated at order $r+1$) as the following Butcher-type tree series:

$$u_k^r(t) = \sum_{T \in \mathcal{V}_k^r} \frac{\Upsilon^p(T)}{S(T)} \Pi(T)(t), \tag{5.8}$$

where $S(T) \in \mathbb{N}$ is the symmetry factor associated to the tree T, $\Upsilon^p(T)$ is the elementary differential which depends on $u_k(0)$ and encodes the coefficient appearing in the iteration of Duhamel's formulation, and $\Pi(T)(t)$ represents a Fourier iterated integral. In their chapter, the authors denote the above expansion (5.8) by $U_k^r(t, u_0)$. Lastly, we give an example of the set $\mathcal{V}_k^r$: for a first-order scheme we fix $r = 0$ and obtain

$\mathcal{V}_k^0 = \{T_0, T_1\}$, with $T_0 =$ , $T_1 =$ 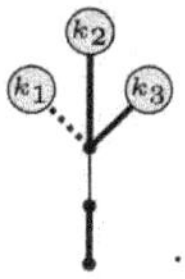 .

The above tree based formalism is inspired by the work [40]. There, this formalism is used to derive a general framework for solving a large class of singular SPDEs in subcritical-regimes, via the theory of Regularity Structures.

[b.] *On the discrete level.* The second step is to approximate each Duhamel integral $\Pi(T)(\tau)$, where τ denotes the time step. The approximation of each Duhamel integral is made using the resonance-based approach detailed previously (see Section 2.1). The discretization is automated via a notion of high and lower parts (2.15) dictated by the resonance structure $R(k)$ of the underlying equation. This automatization is led by the approximation operator

$$\Pi^{r,n} : \mathcal{V}_k^r \to \textit{ approximate iterated integral},$$

which is a resonance-based approximation of Π. Namely, given a decorated tree T, $\Pi^{r,n}(T)$ is a resonance-based approximation of the iterated integral encoded by the tree T. The index n corresponds to a formal a priori assumption on the regularity required on the initial data, and allows to make simple Taylor expansions when sufficient regularity is assumed. The mapping $\Pi^{r,n}$ expresses in Fourier space the numerical scheme after one time step, denoted by $U_k^{n,r}(\tau)$, as the following tree series:

$$U_k^{n,r}(\tau) = \sum_{T \in \mathcal{V}_k^r} \frac{\Upsilon^p(T)}{S(T)} \Pi^{n,r}(T)(\tau). \tag{5.9}$$

We now give the author's definition of the big-O notation to express the local error term.

Definition 9. Let $\Phi^\tau(u_0) = u^1 \approx u(\tau)$ denote the numerical solution at time $t = \tau$. We write

$$u(\tau) - \Phi^\tau(u_0) = \mathcal{O}_{\|\cdot\|}(\tau^m \tilde{\mathcal{L}} u_0), \tag{5.10}$$

if in a suitable norm $\|\cdot\|$, it holds that

$$\|u(\tau) - \Phi^\tau(u_0)\| \leq C(T,d)\tau^m \sup_{0 \leq t \leq \tau} \|q\left(\tilde{\mathcal{L}} u(t)\right)\|, \tag{5.11}$$

for some polynomial q, differential operator $\tilde{\mathcal{L}}$, and constant C independent of τ.

If (5.10) holds, it is said that the numerical solution u^1 approximates the exact solution $u(t)$ at time $t = \tau$ with a local error of order $\mathcal{O}_{\|\cdot\|}(\tau^m \tilde{\mathcal{L}} u_0)$.

The main theorem of the work [38] is the following formal local error expansion, which sums all the local error made by the approximation of each iterated integral:

$$U_k^{n,r}(\tau) - u_k^r(\tau) = \sum_{T \in \mathcal{V}_k^r} \mathcal{O}_{\|\cdot\|}\left(\tau^{r+2} \mathcal{L}_{\text{low}}^r(T,n) \Upsilon^p(T)\right), \tag{5.12}$$

where $\mathcal{L}_{\text{low}}^r$ involves all lower-order frequency interactions neglected during the resonance-based approximation. The global order of the scheme is $r+1$, and hence the local error is of order $r+2$.

Example 10 (First-order scheme for NLS). For the first-order low-regularity approximation to the NLS equation, we fix $r = 0$ and $n = 1$ in equation (5.9). The scheme after one time step is given by

$$U_k^{1,0}(\tau) = e^{-i\tau k^2} u_k(0) - i\tau e^{-i\tau k^2} \sum_{k=-k_1+k_2+k_3} \bar{u}_{k_1}(0) u_{k_2}(0) u_{k_3}(0) \varphi_1(2i\tau k_1^2),$$

and coincides with the first-order resonance-based scheme (2.20). The only approximation made is on the integral $\Pi(T_1)$, where T_1 is the tree given by the second term in (5.7). Moreover, $\mathcal{L}^0_{\text{low}}(T_1, n = 1)$ is given in Fourier by $R_{\text{low}} = -2k_1(k_2 + k_3) + 2k_2k_3$, and can be mapped in physical space to (2.21). In [38, Corollary 5.1] the authors write that the local error is of order $\mathcal{O}_{\|\cdot\|}(\tau^2|\nabla|u)$, meaning that one obtains a local error of order τ^2 in a suitable norm $\|\cdot\|$, if one can bound the term (2.21) in this norm. The choice of the norm $\|\cdot\|$ dictates the regularity assumptions needed on the solution to obtain global convergence, see Remark 11. Fixing the regularity index n to one, allows to construct a scheme whose local error only involves differential operators of order one. By taking $n = 2$, one recovers the classical Euler exponential method (2.6), whose local error is of order $\mathcal{O}_{\|\cdot\|}(\tau^2\Delta u)$, involving differential operators of order two.

Obtaining the formal local error term (5.12) is an important first step towards obtaining convergence of the scheme. However, to obtain a global convergence result, one needs to fix the space and norm $\|\cdot\|$ in which to work with, rigorously bound the local error term in this norm (as in (5.11)) and couple this with a stability argument to conclude. Deducing the global error bounds from the local error terms is in general nontrivial for PDE problems. We cite for example the works [11, 12, 21, 22] which deal with this rigorous analysis in the case of the NLS equation.

Remark 11 (Rigorous regularity assumptions). The choice of norm taken in Definition 9 is important and determines the regularity needed to bound the local error term (5.11), and hence for the convergence of the scheme. We illustrate this on the first-order approximation to the NLS equation, by taking two different choices of norms and quantifying the regularity assumptions thereby required.

When measuring in L^2-norm: one needs to bound the local error term (2.21) in L^2, which by using classical Sobolev embedding requires $u_0 \in H^{1+\frac{d}{4}}$. Indeed, for each of the terms in (2.21), we have that for $d \leq 3$:

$$\|u\nabla u \cdot \nabla u\|_{L^2} \leq C_d\|u\|_{L^\infty}\|\nabla u\|^2_{L^4} \leq C_d\|u\|_{H^{\frac{d}{2}+\epsilon}}\|\nabla u\|^2_{H^{\frac{d}{4}}} \leq C_d\|u\|^3_{H^{1+\frac{d}{4}}},$$

where we used the Sobolev embeddings $H^{\frac{d}{4}} \hookrightarrow L^4$ and $H^\sigma \hookrightarrow L^\infty$, $\sigma > d/2$.

We make the important point that even though the formal local error term in [38, Corollary 5.1] reads $\mathcal{O}_{\|\cdot\|}(\tau^2|\nabla|u)$, one cannot bound this local error term in L^2 while only asking for H^1-solutions. This is due to the polynomial q appearing in (5.11).

When measuring in H^σ, $\sigma > d/2$: using the fact that the space is an algebra we can bound the local error term (2.21) for $u_0 \in H^{\sigma+1}$:

$$\|u\nabla u \cdot \nabla u\|_{H^\sigma} \leq C_d \|u\|_{H^\sigma} \|\nabla u\|^2_{H^\sigma} \leq C_d \|u\|^3_{H^{\sigma+1}}.$$

This analysis is made more generally in H^σ-norm for $\sigma \geq 0$ in [22], where first- and second-order convergence results are obtained.

We finish by discussing a limitation of the work [38]. Given that their algorithm heavily relies on expansions made in Fourier space, the method is restricted to spatial domains which are periodic and to dispersive equations with polynomial nonlinearities. The work [35] overcomes this limitation and is presented next. The authors introduce a nested commutator structure to pull out the oscillatory phases, which replaces and generalizes the Fourier-based constructions. This work extends the tools presented in Section 5 to higher-order.

5.1.2. *Higher-order low-regularity integrators*

The work [35] uses an algebraic framework, to systematically derive low-regularity integrators on the general domains of Section 5. To illustrate, we consider the approximation, up to arbitrary order, to the following class of evolution equations:

$$\begin{cases} \partial_t u(t,x) - \mathcal{L} u(t,x) = \sum_\ell f_\ell\big(u(t,x), \overline{u}(t,x)\big) V_\ell(x), \\ u(0,x) = u_0(x), \quad x \in \Omega, \quad \Omega \subseteq \mathbb{R}^d. \end{cases} \tag{5.13}$$

Here, $\mathcal{L}$ and f_ℓ satisfy the same hypotheses as $\mathcal{L}$ and f in equation (5.1) of Section 5. We recognize the GP equation (5.3) in the case where $V_0 = 1, V_1 = V,\ f_0(u, \bar{u}) = \bar{u}u^2,\ f_1(u, \bar{u}) = u$. The above class of evolution equations is more general than both (1.2) and (5.1).

In order to obtain higher-order approximations, we need to consider Duhamel's iterates (5.4) where we replace the nonlinearity $p(u)$ by $\sum_\ell f_\ell(u, \overline{u}) V_\ell$. We obtain their scheme by discretizing each iterated integral, as detailed below.

(a) *On the continuous level.* This time one needs to encode by decorated trees the Duhamel iterates (5.4) in the physical variables. This generalizes the decorated trees in [38] which relied on Fourier-based expansions, to consider more general domains and equations. We let the brown edge encode the integral of the linear flow:

$$\int_0^t e^{(t-s)\mathcal{L}} \cdots ds = \mathbf{|}\,.$$

Each node of the tree encodes a term in the nonlinearity, and the decoration on the ℓ-th node encodes the potential V_ℓ (and possible monomials). For example, for the GP equation there are two nodes corresponding to the two nonlinear terms ($f_\ell,\ \ell = \{0,1\}$):

$$\circ \equiv (e^{is\Delta}u_0^2)(e^{-is\Delta}\overline{u_0}), \qquad \bullet \equiv (e^{is\Delta}u_0)V.$$

Incoming edges to a node induce derivatives on f_ℓ, allowing to encode Duhamel iterates with more than one integral in time. This allows to iteratively define a set of trees $\mathcal{T}^r$ encoding – through the mapping Π – each Duhamel iterate consisting of $r+1$ integrals in time. We can thus express $u(t)$ truncated at order $r+1$, as the tree series expansion:

$$u^r(t) = e^{t\mathcal{L}}u_0 + \sum_{T\in\mathcal{T}^r} \Pi(T)(t).$$

We note that the symmetry factor $S(T)$ and elementary differential $\Upsilon^p(T)$ previously appearing in (5.8), are directly included in $\Pi(T)$.

(b) *On the discrete level.* The authors define a new operator Π_A^r mapping a decorated tree in $\mathcal{T}^r$ to an *approximate Duhamel iterate.* The map Π_A^r is a discretized version of Π. The desired order of the scheme is $r+1$ and A is a given domain of regularity of u_0 and $(V_\ell)_\ell$, reflecting the a priori regularity assumed on the initial data and the potentials. The role of A was played by the index n in the periodic setting, see (5.9). The operator Π_A^r discretizes the Duhamel iterates at low-regularity by filtering out the dominant parts, integrating them exactly, and approximating the lower-order parts through commutator based expansions. The numerical scheme U_A^r, after one time step, is given by:

$$U_A^r(\tau) = e^{\tau\mathcal{L}}u_0 + \sum_{T\in\mathcal{T}^r} \Pi_A^r(T)(\tau).$$

For the sake of clarity, we have not included in the above discussion a technical preliminary step, which extracts the linear flow $e^{t\mathcal{L}}u_0$ from inside the nonlinearity $p(e^{t\mathcal{L}}u_0)$. This step gives a unified form of the Duhamel iterates, allowing to encode them using decorated trees.

The main theorem of [35] is a formal local error expansion, measuring the error made when approximating each Duhamel iterate over one time step.

Theorem 12. *([35, Theorem 4.3] in a simplified notation)*

$$U_A^r(\tau) - u^r(\tau) = \sum_{T \in \mathcal{T}^r} \mathcal{O}_{\|\cdot\|}\left(\tau^{r+2}\mathcal{L}^r_{low}(T, A)\right), \tag{5.14}$$

where the operator $\mathcal{L}^r_{low}(T, A)$ embeds the necessary regularity on the solution.

The above accounts for the local error made by the approximation in step *b*. In the end, one needs to sum the above error terms with the error from the truncation made in step *a*. If the nonlinearity considered does not involve derivatives, step *b* is the dominant one and dictates the needed regularity.

We stress the fact that these higher-order techniques only guarantee a formal local error bound as above, where the big-O notation is to be understood in the sense of Definition 9. In order to obtain convergence of the scheme, a separate and careful analysis needs to be made: firstly for estimating the local error in an appropriate functional space, and secondly to obtain stability of the scheme.

For an example of this algorithmic approach, we refer to [35, Section 5.1] where classical and low-regularity first- and second-order integrators for the GP equation are obtained, together with their formal local error expansion. To obtain a rigorous convergence result of these schemes, we once again refer to the results of [22].

We finish this section with a short historic note. The idea of using a tree-series based expansions to express the numerical scheme was heavily studied in the context of ODEs in the 70s by the pioneering works of [41, 42]. More precisely, it was used for the characterization of Runge-Kutta methods via Butcher series (B-series), see the reviews [1, 43]. The algebraic framework used at this time was a foundation first for Gubinelli's approach to Rough Path Theory [44, 45], and afterwards to Hairer's theory of regularity structures [40], a generalization of Rough Paths. Nowadays, these ideas are making their way back to the study of numerical schemes, but for PDEs instead of ODEs.

Remark 13 (Randomized initial data). In this section we discussed the design to higher order of new families of numerical schemes for *deterministic* PDEs, suited for *low-regularity* initial data. In the work [46] we enter a probabilistic setting by considering a *randomized* initial condition. We construct approximations of arbitrarily high order to the second moment of the Fourier coefficient of the solution, based upon the tree series expansion (5.7), Feynman diagrams and Wick's formula.

6. Symmetric resonance-based schemes

In this section we discuss topic 4, on structure preservation.

6.1. *A symmetric scheme for the nonlinear Schrödinger equation, suited for non-smooth solutions*

We start by presenting the work [12] on the NLS equation (1.1). The NLS equation is time reversible, meaning that $\overline{u(-t,x)}$ is again solution of (1.1), and preserves the mass

$$\|u(t)\|_{L^2}^2 = \int |u(t,x)|^2 dx = \|u_0\|_{L^2}^2,$$

and the energy

$$E(t) = \frac{1}{2}\int |\nabla u(t,x)|^2 dx + \frac{1}{4}\int |u(t,x)|^4 dx = E(0)$$

of the solution over its interval of existence in time.

This work builds on the observation that low regularity schemes discussed in earlier sections do not preserve the invariants of the underlying equation, such as the mass and energy for the NLS equation, see for example Figure 2.

In the work [12] the construction and rigorous analysis of a novel scheme with strong geometric structure at low regularity is given, allowing for a reliable symmetric approximation to the NLS equation. Namely, the scheme preserves the time-reversibility of the equation and numerical simulations show good near-preservation of the mass and energy over long times. Moreover, we witness experimentally that the new symmetric low-regularity scheme has better long time near-preservation of the energy than classical splitting schemes. We refer to Figure 2 where we plot the error in the energy over time, first for a smooth initial condition $u_0 \in C^\infty$, then in the case of a rougher initial condition $u_0 \in H^2$. We notice that the first-order resonance-based scheme (pink) does not preserve the energy, even over short times. In Figure 2(a) the Lie and Strang splitting (yellow and blue) approximately preserves the energy up until a point ($t \sim 30$), while in the rougher case of Figure 2(b), the energy explodes after much shorter times. This behavior

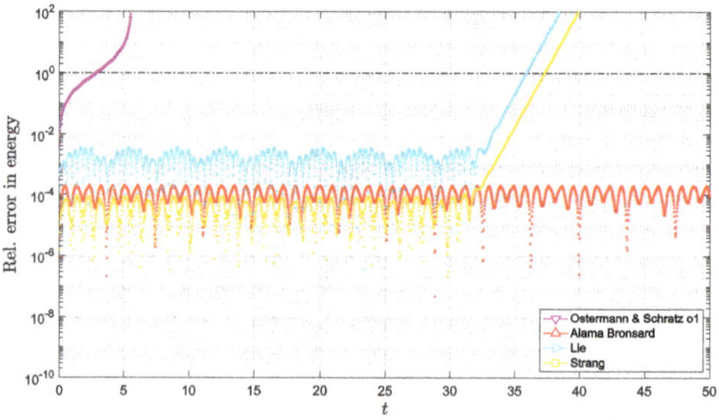

(a) C^∞ data.

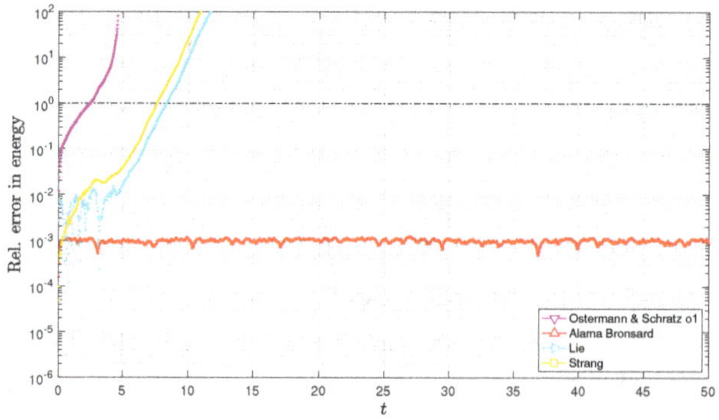

(b) H^2 data.

Fig. 2. Error in the Hamiltonian for C^∞ smooth data and less regular H^2 data. We take $\tau = 0.02$ and $K = 1024$. See the work [36] for further numerical experiments.

is rigorously studied in the work of [2] where the author shows that a CFL condition is necessary to guarantee long-time approximate energy preservation for splitting methods, see also Remark 14. Namely, if we fix the time step τ and increase the highest Fourier frequency K, thereby going from an ODE to the (desired) PDE setting, the energy ceases to be preserved at some point. In comparison, the new symmetric low-regularity scheme (red) does not suffer from a comparable CFL condition, and performs well for both smooth and low-regularity solutions.

Remark 14 (Near energy preservation of splitting methods for the NLS equation under a CFL condition). Splitting methods preserve the L^2 norm and symplectic structure of the solution. For the NLS equation, the symplectic form is given by

$$\sum_{a\in\mathbb{Z}} \mathrm{d}\xi_a \wedge \mathrm{d}\eta_a,$$

where $\xi_a = \hat{u}_a$ and $\eta_a = \overline{\hat{u}_a}$ are the Hamiltonian coordinates, see [2, Section III.1]. However, they do not preserve exactly the energy on the discrete level. Indeed, under suitable assumptions, the numerical approximation can be shown to solve (almost) exactly a *modified* PDE at each time step. This results in the existence of a *modified* energy which is preserved along the numerical flow, where – among other assumptions – for smooth solutions and under a CFL condition, it can be shown that the modified energy is close to the exact energy. More specifically, the CFL condition imposes that the time step size τ has to be chosen such that $\tau \lesssim K^{-2}$ where K denotes the highest frequency in the discretization, see [2]. This step size restriction is not only a theoretical technicality, but also observed in numerical experiments, see Figure 2. We also refer to a number of important works where the near-preservation of the energy over exponentially long times is shown using modulated Fourier expansions (see [47–49]) or normal form techniques ([50–52]).

We now portray the idea behind the construction of this symmetric scheme. We recall that a scheme is said to be *symmetric* or *time reversible* if it remains unchanged under the transformations $\tau \to -\tau$ and $n \leftrightarrow n+1$.

We once again start from Duhamel's integral mapped in Fourier space (2.12), whose integrand is decomposed in (2.14) into the three factors: $e^{isR_{\mathrm{dom}}(k)}e^{isR_{\mathrm{low}}(k)}g(s)$, with $R(k) = R_{\mathrm{dom}}(k) + R_{\mathrm{low}}(k)$ given in (2.15). To construct the first-order resonance-based scheme (2.20) we simply Taylor expanded both $g(s)$ and $e^{isR_{\mathrm{low}}(k)}$ about the left end point: $e^{isR_{\mathrm{low}}(k)}g(s) \approx g(0)$. This however breaks the symmetric structure of the underlying PDE and leads to schemes which are not structure preserving. We thus look for a symmetric approximation of both the non-oscillatory $g(s)$ and lower part $e^{isR_{\mathrm{low}}(k)}$. In the work [12] the following symmetric approximation is chosen: $h(s) \approx h(0)\mathbb{1}_{[0,\tau/2]} + h(\tau)\mathbb{1}_{(\tau/2,\tau]}$, which in our context yields

$$I_\tau = \int_0^\tau e^{isR_{\mathrm{dom}}(k)} \underbrace{e^{isR_{\mathrm{low}}(k)}g(s)}_{\approx g(0)\mathbb{1}_{[0,\tau/2]}+e^{i\tau R_{\mathrm{low}}(k)}g(\tau)\mathbb{1}_{(\tau/2,\tau]}} ds. \tag{6.1}$$

By integrating exactly the dominant part $e^{isR_{\mathrm{dom}}(k)}$, mapping back to physical space and in the untwisted variable, we obtain the symmetric scheme of [12]:

$$
\begin{aligned}
u^{n+1} &= \Phi_{-\tau/2}^{-1} \circ \Phi_{\tau/2}(u^n) \\
&= e^{i\tau\Delta}u^n - i\frac{\tau}{2}e^{i\tau\Delta}\left((u^n)^2\varphi_1(-i\tau\Delta)\overline{u^n}\right) \\
&\quad - i\frac{\tau}{2}\left((u^{n+1})^2\varphi_1(i\tau\Delta)\overline{u^{n+1}}\right),
\end{aligned}
\tag{6.2}
$$

where Φ_τ is the first-order resonance-based scheme (2.20) constructed in the introduction. In fact, the symmetric approximation (6.1) is chosen because it coincides with the symmetrization of the first-order resonance-based scheme Φ_τ. For other choices of symmetric approximations, we refer to Section 6.2.

A notable difference between this symmetric scheme and previous low-regularity schemes is its *implicit* nature. We refer to Figure 3 which shows that the computational cost of the symmetric scheme is comparable to that of the asymmetric second-order resonance-based scheme. Namely, its improved convergence properties make up for the extra cost of solving the implicit system (6.2) at each time step.

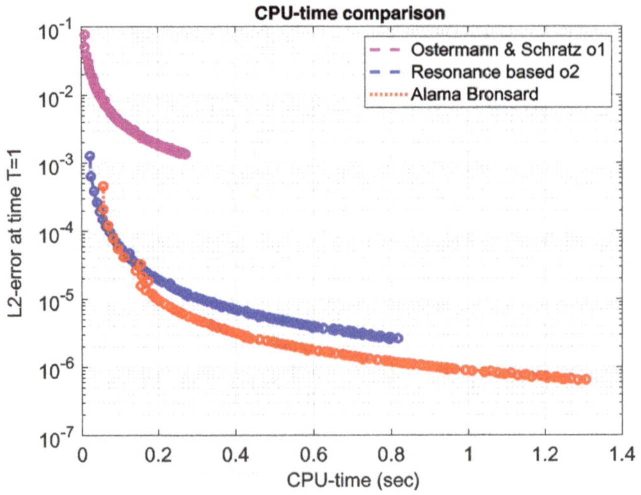

Fig. 3. CPU time comparison for H^1 data. The first-order resonance-based scheme is in pink (Ostermann & Schratz o1) and in blue figures the second-order resonance-based scheme (Resonance based o2) which can be found for example in [38]. The symmetric scheme in red (Alama Bronsard) is given by (6.2).

A desired property of symmetric schemes is that they are of *even order*. This however is only true under sufficient regularity assumptions of the solution. The work [12] focuses on obtaining the minimal regularity assumptions needed for convergence of the scheme from first to second order. The error analysis presented is suited both for periodic domains $\Omega = \mathbb{T}^d$ and smooth bounded domains $\Omega \subset \mathbb{R}^d$ with homogeneous Dirichlet boundary conditions. The author works in the fractional space

$$X^\sigma(\Omega) = \mathcal{D}((-\Delta)^{\sigma/2}), \quad ||u||^2_{X^\sigma} = ||u||^2_{L^2} + ||(-\Delta)^{\sigma/2}u||^2_{L^2}.$$

In particular $X^\sigma(\mathbb{T}^d) = H^\sigma(\mathbb{T}^d)$, and in the more subtle case of a smooth bounded domain, the space $X^\sigma(\Omega)$ embeds both the regularity and the necessary boundary conditions. In this case, one is led to characterize domains of fractional powers of the Dirichlet Laplacian, via complex interpolation spaces, see [12, Section 2.2]. The convergence result then reads:

$$||u(n\tau) - u^n||_{L^2} \leq C_T \left(\sup_{[0,T]} ||u(t)||_{X^\alpha} \right) \tau^{1+\gamma}, \quad 0 \leq n\tau \leq T, \quad \gamma \in [0,1],$$

where the regularity α is minimized through well-chosen bilinear estimates. The admissible values of α and γ are represented by the blue zone on the right picture. The detailed theorems and the bilinear estimates are given in [12, Section 1.2 and Section 2] respectively.

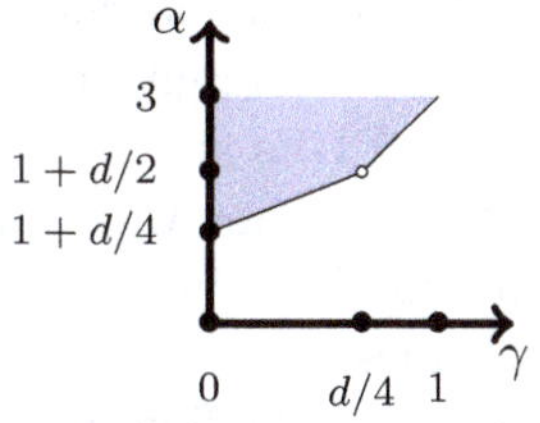

We refer to Figure 1 where convergence plots are made for H^1 and H^2 initial data. We observe that the symmetric scheme (red) has a better error constant than the previous resonance-based scheme (pink), the exponential integrator (green), and the Lie splitting method (light blue).

We give a brief overview of current *local in time* results that build new bridges between structure preserving methods and low-regularity approaches. In [24], a first-order Fourier integrator is introduced for the NLS equation (1.1) set on the 1-dimensional torus $\mathbb{T}$, which almost conserves the mass. For the cubic Klein-Gordon equation set on $\mathbb{T}$, an explicit symmetric three time-step low-regularity integrator was obtained by [53], using Gautschi-type methods. The authors of [54] introduce a framework for building novel symplectic resonance-based integrators, encompassing both the KdV equation and the 1-dimensional NLS equation. A symmetrized resonance-based integrator for the Schrödinger map (SM) is obtained in [55], where they relate the SM flow to the NLS equation set on 1-d torus $\mathbb{T}$ via the Hasimoto transform. Finally, the recent work [56] establishes a second-order explicit and symmetric three-time step method for the GP equation (5.3) with a low-regularity potential V, using Gautchi-type exponential wave integrators.

Remark 15 (From short to long time scales). All of these convergence results allow to rigorously prove the reliability of a numerical discretization over short times $[0, T]$, as the error constant C_T usually depends exponentially on the final time T. Pushing the analysis to obtain error estimates over longer times is thus a highly nontrivial problem.

An important step was made by the authors of [57]. Using scattering theory in order to obtain quantitative time decay estimates, they show uniform in time error estimates for the NLS equation on the full space $\mathbb{R}^d$, for a Lie splitting discretization. Their convergence analysis is, however, limited to $\mathbb{R}^d$ as it heavily relies on dispersive effects, which do not hold on the torus $\mathbb{T}^d$ or more generally on compact domains. The step from the full space to the compact setting is – as in the continuous setting – nontrivial since in the latter dispersion does not translate into decay. Nevertheless, compact domains are computationally very interesting as spatial discretizations of nonlinear PDEs are in general performed in truncated domains.

This limitation was addressed for the first time in [58] for certain nonlinear and nonlocal integrable equations. Specifically, they developed numerical schemes that converge on the torus $\mathbb{T}$, with an error constant that grows linearly with the final time T. This improvement is achieved by leveraging the integrable structure of the equation. Unlike on the full space $\mathbb{R}$, the error on the torus is expected to grow linearly over time. In this sense, a linear dependency on T represents the optimal error growth in the periodic setting.

We note that the above results are not constrained to small initial data $\|u_0\| \leq \epsilon$. For readers interested in this case, the treatment of small initial data using low-regularity techniques can be found in [59].

However, all these results are yet again tailored to the particular structure of the equation, and their higher-order extensions are not optimal in the sense of regularity, see Remark 17. This motivates the study of systematic constructions of symmetric resonance based schemes that we discuss next.

6.2. *Symmetric higher-order extensions*

A natural starting point is to consider symmetrizing existing asymmetric resonance based schemes. Typically, given Φ_τ an asymmetric resonance based scheme of order p, we expect the symmetrization with its adjoint

$$\widehat{\Phi}_{\tau/2} \circ \Phi_{\tau/2},$$

to be a p-th order method with the same regularity requirements as for Φ_τ, and when p is odd, one can show through appropriate additional regularity assumptions that the scheme is of order $p+1$, see [12]. This yields a *subclass* of high-order symmetric resonance-based schemes, built upon the asymmetric resonance-based schemes of [38]. Nonetheless, myriads of other possible symmetric approximations exist. The goal of the work [36] is to offer a general characterization for symmetric schemes, while putting up front a new subclass of high-order symmetric schemes based upon mid-point approximations.

The work [36] extends the resonance based decorated trees approach [38] to a richer framework by exploring different ways of iterating Duhamel's formula, capturing the dominant parts while interpolating the lower parts of the resonances in a symmetric manner. This gives a range of new numerical schemes with more degrees of freedom than the original framework from [38]. The new framework in [36] allows us to recover previously constructed low-regularity symmetric schemes, such as the scheme (6.2), as well as new symmetric low-regularity schemes which includes the previously mentioned mid-point based approximations, which we detail next.

We first recall Duhamel's formula (2.1), which, in the twisted variables (2.8) and mapped in Fourier, corresponds to

$$v_k(t) = v_k(0) + \int_0^t e^{-s_1 P(k)} p_k \left(e^{s_1 P(k)} v_k(s_1) \right) ds_1. \tag{6.3}$$

Previously, to obtain higher-order schemes we iterated Duhamel's formula about the linear flow $e^{s_j \mathcal{L}} u_0$, yielding (5.4) and (5.6) in Fourier space. In the twisted variable, this amounts to iterating the above equation (6.3) about the left-end point $v_k(0)$. Namely, we substitute $v_k(s_1) = v_k(0) + \int_0^{s_1} \cdots ds_2$ in the above equation, then use a Taylor expansion in p_k about $e^{s_1 P(k)} v_k(0)$.

In the previous Section 6.1, we saw that making this asymmetric approximation about the left end point $v_k(s) \approx v_k(0)$ (or equivalently $g(s) \approx g(0)$ in (6.1)) leads to schemes which are *not structure preserving*. The solution we propose is to make a symmetric approximation of the non-oscillatory part $g(s)$ (and hence $v_k(s)$), such as the approximation

$v_k(s) \approx v_k(0)\mathbb{1}_{[0,\tau/2]} + v_k(\tau)\mathbb{1}_{(\tau/2,\tau]}$, which puts equal weights on both end points $s = 0$ and $s = \tau$. Hence, the first step towards obtaining higher-order symmetric schemes is to iterate Duhamel's formula in $v_k(s_j)$ in a symmetric fashion, by putting equal weights on Duhamel's formulas centered about 0 and τ:

$$v_k(s_j) = v_k(0) + \int_0^{s_j} e^{-s_{j+1}P(k)} p_k\left(e^{s_{j+1}P(k)} v_k(s_{j+1})\right) ds_{j+1},$$

$$v_k(s_j) = v_k(\tau) + \int_{s_j}^{\tau} e^{-s_{j+1}P(k)} p_k\left(e^{s_{j+1}P(k)} v_k(s_{j+1})\right) ds_{j+1}.$$

In [36] we first take the mid-point rule as our symmetric approximation:

$$v_k(s_j) = \frac{v_k(0) + v_k(\tau)}{2} + O(\tau), \quad s_j \in [0,\tau]. \tag{6.4}$$

We let $r+1$ denote the order of the scheme. We iterate the above equation (6.4) inside Duhamel's formula (6.3) and obtain an expansion of the continuous solution (written in the untwisted variables) and up to order $r+1$ as a tree series

$$u^r_{\text{mid},k}(t) = \sum_{T\in\mathcal{V}^r_k} \frac{\Upsilon^p_{\text{mid}}(T)}{S(T)} \Pi_{\text{mid}}(T)(t). \tag{6.5}$$

The above series is of the same form as the tree-series expansion (5.8), the only difference lies in the definition of the character Π and coefficient Υ^p. Here, we encode the symmetric iterations of Duhamel's formula in Π_{mid} and Υ^p_{mid}, whereas Π and Υ^p encoded Duhamel's iterates centered about zero (see (5.6) or (6.3)). Note that in [36] everything is directly written in the untwisted variables. It is however much more intuitive to start by working in the twisted variables, and to make the decomposition (2.12) and (2.14), as presented above.

The next step is to discretize at low-regularity each Duhamel iterate $\Pi_{\text{mid}}(T)(\tau)$ in a symmetric fashion. This is to be compared with Section 6.1: up to this point we have approximated $g(s)$ in (6.1), and we are now ready to approximate the lower part $e^{isR_{\text{low}}(k)}$. We introduce the operator $\Pi^{n,r}_{\text{mid}}$, which based on the frequency decomposition $R(k)$ (and on the a priori degree of regularity of the solution n), gives a symmetric low-regularity approximation of the remaining oscillatory factors, based on polynomial interpolation. After one time step the scheme takes the form

$$U^{n,r}_{\text{mid},k}(\tau) = \sum_{T\in\mathcal{V}^r_k} \frac{\Upsilon^p_{\text{mid}}(T)}{S(T)} \Pi^{n,r}_{\text{mid}}(T)(\tau), \tag{6.6}$$

together with the formal local error expansion:

$$U^{n,r}_{\text{mid},k}(\tau) - u^r_{\text{mid},k}(\tau) = \sum_{T \in \mathcal{V}^r_k} \mathcal{O}_{\|\cdot\|}\left(\tau^{r+2}\mathcal{L}^r_{\text{low}}(T,n)\Upsilon^p_{\text{mid}}(T)\right). \tag{6.7}$$

This is to be compared with the tree series (5.9) and local error (5.12).

Throughout this iterative procedure of building the scheme, we choose the mid-point rule as our symmetric approximation, and obtain the class (6.6). This subclass differs from the classical subclass of symmetrized schemes $\widehat{\Phi}_{\tau/2} \circ \Phi_{\tau/2}$, and has the advantage that the derivation of these new schemes involve fewer terms. For example, the scheme obtained by composing the second-order (non-symmetric) resonance-based scheme for NLS with its adjoint, yields a scheme with more terms, including the second-order term $\frac{\tau^2}{8}(|u^{n+1}|^4 u^{n+1} - e^{i\tau\Delta}|u^n|^4 u^n)$, which does not appear in our new scheme proposed in Proposition 4.13 (or equivalently equation (4.18)) thanks to symmetry cancellations during the resonance-based mid-point approximation.

We are also interested in obtaining a more general class of symmetric approximations encompassing the class (6.6), the previously built symmetric integrators (such as (6.2)), as well as new symmetric integrators. In [36, Section 3.2] we derive a *forest formula*, which characterizes a large class of symmetric schemes:

$$\begin{aligned} u^{\ell+1}_k = e^{i\tau\mathcal{L}} u^\ell_k + e^{i\tau\mathcal{L}} \sum_{T \in \tilde{\mathcal{V}}^r_k} \sum_{\mathbf{a} \in [0,1]^{\tilde{E}_T}} \sum_{\chi \in \{0,1\}^{L_T}} \sum_{T_0 \cdot T_1 \cdots T_m \subset T} C_T \\ b_{\mathbf{a},\chi,T,T_0\cdot\ldots\cdot T_m}(\tau, i\tau\mathscr{F}_{\text{dom}}(T_j), j \in \{0,\ldots,m\}) \\ \prod_{e \in \tilde{E}_{T_j}} e^{i\tau a_e \mathscr{F}_{\text{low}}(T^e_j)} \frac{\Upsilon^p_\chi(T)(u^{\ell+\chi_v}_{k_v}, v \in L_T, \tau)}{S(T)}. \end{aligned} \tag{6.8}$$

We derive a general formula (6.8) parametrizing resonance-based and classical integrators and deduce conditions on the coefficients of this formula to obtain a symmetric scheme, see [36, Proposition 4.7 and Section 4.2]. While the forest formula (6.8) comprises a large class of integrators, it is not defined in a recursive manner, hence obtaining a closed formula for the local error terms (such as (6.7)) is not obvious. The advantage of fixing a symmetric approximation (such as the mid-point rule (6.4)) and constructing the iterative procedure (6.5) on each tree T, is that one has a derivation of the local error terms (6.7) at hand.

Obtaining the above local error term is an important first step towards obtaining convergence of the scheme. However, to obtain a global convergence result, one needs to fix the space and norm $\|\cdot\|$ in which to work with, rigorously bound the local error term in this norm (as in (5.11)) and couple this with a stability argument to conclude.

We refer to [36, Section 5] where new symmetric integrators are tested numerically, and show good near-preservation of the energy without CFL conditions.

Remark 16 (Symmetrizing by composition with its adjoint). Composing a non-symmetric resonance-based method Φ_τ with its adjoint $\widehat{\Phi}_\tau$, yields a subclass of high order symmetric resonance-based schemes. In particular, the subclass of schemes written as the composition $\widehat{\Phi}_{\tau/2} \circ \Phi_{\tau/2}$, can be recovered from the forest formula presented in equation (6.8). On the other hand, the schemes of the form $\Phi_{\tau/2} \circ \widehat{\Phi}_{\tau/2}$ are implicit two time-step methods, which are not included in the framework of [36]: we only deal with one time-step methods, written as an equation on u^{n+1} and u^n. For two time-step methods, involving u^n, $u^{n+1/2}$ and u^{n+1}, one expects to obtain explicit schemes instead of implicit ones, in the case of the Klein-Gordon equation we refer to [53] and to [60] for the NLS equation).

Remark 17 (Optimal schemes in regards to the regularity conditions required). One needs to be careful that the symmetric schemes thereby obtained don't require more regularity than the (non-symmetric) resonance-based schemes of [38]. Indeed, following the error analysis result in [12], composing the first-order resonance-based scheme (2.20) with its adjoint, yields the symmetrized scheme (6.2) of the form $\widehat{\Phi}_{\tau/2} \circ \Phi_{\tau/2}$, which is of first and second order respectively for solutions in $H^{\sigma+1}$ and $H^{\sigma+3}$, where H^σ, $\sigma > d/2$, is the Sobolev norm in which we measure the error. In contrast, previous (non-symmetric) resonance-based schemes require $H^{\sigma+1}$ and $H^{\sigma+2}$ solutions for convergence at the first and second order respectively, and we expect the p-th order scheme to converge at order p for solutions in $H^{\sigma+p}$. Thus, the symmetrized scheme $\widehat{\Phi}_{\tau/2} \circ \Phi_{\tau/2}$ is optimal as a first-order scheme, but a different second-order scheme needs to be taken for it to be optimal in regard to the regularity assumptions. We note that this also holds for the mid-point scheme (6.6). To obtain an optimal symmetric second-order scheme we push the resonance-based analysis and exploit the symmetry to obtain cancellation of higher-order terms yielding in [36, Proposition 4.13 and Definition 3.13] the second-order scheme and the higher-order midpoint resonance-based schemes respectively, which only require p additional derivatives on the solution.

Remark 18 (Symmetric versus symplectic schemes). The idea behind the works [12, 36] is heavily based upon the decompositions (2.14) and (2.15). We first iterate Duhamel's formula in the nonlinearity in a symmetric fashion, give a symmetric approximation to the lower-order parts and integrate the dominant part exactly. This yields symmetric schemes. To obtain *symplectic resonance-based schemes*, a different decomposition of the oscillatory terms $e^{isR(k)}$ must be done to ensure that the approximation originates in another Hamiltonian system with the same symplectic structure and conservation of normalization. A suitable decomposition of the frequencies $R(k)$ was obtained by the authors of [54] in the case of the KdV equation and the 1-dimensional NLS equation.

References

[1] E. Hairer, C. Lubich, and G. Wanner, *Geometric numerical integration.* vol. 31, *Springer Series in Computational Mathematics*, Springer, Heidelberg (2010). Structure-preserving algorithms for ordinary differential equations, Reprint of the second (2006) edition.

[2] E. Faou, *Geometric numerical integration and Schrödinger equations.* vol. 15, European Mathematical Society (2012).

[3] C. Besse, A relaxation scheme for the nonlinear Schrödinger equation, *SIAM J. Numer. Anal.* **42**(3), 934–952 (2004).

[4] C. Besse, S. Descombes, G. Dujardin, and I. Lacroix-Violet, Energy-preserving methods for nonlinear Schrödinger equations, *IMA J. Numer. Anal.* **41**(1), 618–653 (2021).

[5] J. Sanz-Serna, Methods for the numerical solution of the nonlinear Schrödinger equation, *Math. Comp.* **43**(167), 21–27 (1984).

[6] C. Lubich, On splitting methods for Schrödinger-Poisson and cubic nonlinear Schrödinger equations, *Math. Comp.* **77**(264), 2141–2153 (2008).

[7] C. Besse, B. Bidégaray, and S. Descombes, Order estimates in time of splitting methods for the nonlinear Schrödinger equation, *SIAM J. Numer. Anal.* **40**(1), 26–40 (2002).

[8] P. Henning and D. Peterseim, Crank-Nicolson Galerkin approximations to nonlinear Schrödinger equations with rough potentials, *Math. Models Methods Appl. Sci.* **27**(11), 2147–2184 (2017).

[9] M. Hochbruck and A. Ostermann, Exponential integrators, *Acta Numer.* **19**, 209–286 (2010).

[10] M. Hofmanová and K. Schratz, An oscillatory integrator for the KdV equation, *Numer. Math.* **136**, 1117–1137 (2017).

[11] A. Ostermann and K. Schratz, Low regularity exponential-type integrators for semilinear Schrödinger equations, *Found. Comput. Math.* **18**, 731–755 (2018).

[12] Y. Alama Bronsard, A symmetric low-regularity integrator for the nonlinear Schrödinger equation, *IMA J. Numer. Anal.* **44**, 3648–3682 (2024).

[13] J. Bourgain, Fourier transform restriction phenomena for certain lattice subsets and applications to nonlinear evolution equations. I. Schrödinger equations, *Geom. Funct. Anal.* **3**(2), 107–156 (1993).
[14] M. Gubinelli, Rough solutions for the periodic Korteweg–de Vries equation, *Commun. Pure Appl. Anal.* **11**(2), 709–733 (2012).
[15] J. D. Lawson, Generalized Runge-Kutta processes for stable systems with large lipschitz constants, *SIAM J. Numer. Anal.* **4**(3), 372–380 (1967).
[16] M. Hochbruck, J. Leibold, and A. Ostermann, On the convergence of Lawson methods for semilinear stiff problems, *Numer. Math.* **145**(3), 553–580 (2020).
[17] R. R. Coifman and Y. Meyer, On commutators of singular integrals and bilinear singular integrals, *Trans. Amer. Math. Soc.* **212**, 315–331 (1975).
[18] R. I. McLachlan and G. R. W. Quispel, Splitting methods, *Acta Numer.* **11**, 341–434 (2002).
[19] N. Burq, P. Gérard, and N. Tzvetkov, Strichartz inequalities and the nonlinear Schrödinger equation on compact manifolds, *Amer. J. Math.* **126**(3), 569–605 (2004).
[20] T. Tao, *Nonlinear dispersive equations: local and global analysis.* vol. 106, Amer. Math. Soc., Providence RI (2006).
[21] A. Ostermann, Y. Wu, and F. Yao, A second-order low-regularity integrator for the nonlinear Schrödinger equation, *Adv. Contin. Discrete Models.* **23**, 14 (2022).
[22] Y. Alama Bronsard, Error analysis of a class of semi-discrete schemes for solving the Gross-Pitaevskii equation at low regularity, *J. Comput. Appl. Math.* **418**, Paper No. 114632, 19 (2023).
[23] A. Ostermann, F. Rousset, and K. Schratz, Fourier integrator for periodic NLS: low regularity estimates via discrete Bourgain spaces, *J. Eur. Math. Soc.* **25**(10), 3913–3952 (2022).
[24] Y. Wu and F. Yao, A first-order Fourier integrator for the nonlinear Schrödinger equation on $\mathbb{T}$ without loss of regularity, *Math. Comp.* **91**(335), 1213–1235 (2022).
[25] B. Li and Y. Wu, A fully discrete low-regularity integrator for the 1d periodic cubic nonlinear Schrödinger equation, *Numer. Math.* **149**(1), 151–183 (2021).
[26] A. Ostermann and F. Yao, A fully discrete low-regularity integrator for the nonlinear Schrödinger equation, *J. Sci. Comput.* **91**(1), 9 (2022).
[27] J. Cao, B. Li, and Y. Lin, A new second-order low-regularity integrator for the cubic nonlinear Schrödinger equation, *IMA J. Numer. Anal.* **44**(3), 1313–1345 (2024).
[28] L. Gauckler, Convergence of a split-step Hermite method for the Gross-Pitaevskii equation, *IMA J. Numer. Anal.* **31**(2), 396–415 (2011).
[29] M. Thalhammer, Convergence analysis of high-order time-splitting pseudospectral methods for nonlinear Schrödinger equations, *SIAM J. Numer. Anal.* **50**(6), 3231–3258 (2012).
[30] J. Eilinghoff, R. Schnaubelt, and K. Schratz, Fractional error estimates of splitting schemes for the nonlinear Schrödinger equation, *J. Math. Anal. Appl.* **442**(2), 740–760 (2016).

[31] G. Dujardin, Exponential Runge-Kutta methods for the Schrödinger equation, *Appl. Numer. Math.* **59**(8), 1839–1857 (2009).

[32] A. Ostermann, F. Rousset, and K. Schratz, Error estimates at low regularity of splitting schemes for NLS, *Math. Comp.* **91**(333), 169–182 (2022).

[33] L. Ji and A. Ostermann, Low regularity error estimates for high dimensional nonlinear Schrödinger equations, *arXiv:2312.11071* (2023).

[34] X. Antoine, W. Bao, and C. Besse, Computational methods for the dynamics of the nonlinear Schrödinger/Gross-Pitaevskii equations, *Comput. Phys. Commun.* **184**(12), 2621–2633 (2013).

[35] Y. Alama Bronsard, Y. Bruned, and K. Schratz, Low regularity integrators via decorated trees, *arXiv:2202.01171* (2022).

[36] Y. Alama Bronsard, Y. Bruned, G. Maierhofer, and K. Schratz, Symmetric resonance based integrators and forest formulae, *arXiv:2305.16737* (2023).

[37] F. Rousset and K. Schratz, A general framework of low regularity integrators, *SIAM J. Numer. Anal.* **59**(3), 1735–1768 (2021).

[38] Y. Bruned and K. Schratz, Resonance-based schemes for dispersive equations via decorated trees, *Forum Math. PI.* **10**, e2 (2022).

[39] A. Connes and D. Kreimer. Hopf algebras, renormalization and noncommutative geometry. In *Quantum field theory: perspective and prospective*, pp. 59–109. Springer (1999).

[40] M. Hairer, A theory of regularity structures, *Invent. Math.* **198**(2), 269–504 (2014).

[41] J. C. Butcher, An algebraic theory of integration methods, *Math. Comp.* **26** (117), 79–106 (1972).

[42] E. Hairer and G. Wanner, On the Butcher group and general multi-value methods, *Computing (Arch. Elektron. Rechnen).* **13**(1), 1–15 (1974).

[43] J. C. Butcher, *Numerical methods for ordinary differential equations*, third edn. John Wiley & Sons, Ltd., Chichester (2016). With a foreword by J. M. Sanz-Serna.

[44] M. Gubinelli, Controlling rough paths, *J. Funct. Anal.* **216**(1), 86–140 (2004).

[45] M. Gubinelli, Ramification of rough paths, *J. Differ. Equ.* **248**(4), 693–721 (2010).

[46] Y. Alama Bronsard, Y. Bruned, and K. Schratz, Approximations of dispersive PDEs in the presence of low-regularity randomness, *Found. Comput. Math.* **24**, 1819–1869 (2024).

[47] D. Cohen, E. Hairer, and C. Lubich, Conservation of energy, momentum and actions in numerical discretizations of non-linear wave equations, *Numer. Math.* **110**(2), 113–143 (2008).

[48] L. Gauckler and C. Lubich, Splitting integrators for nonlinear Schrödinger equations over long times, *Found. Comput. Math.* **10**(3), 275–302 (2010).

[49] L. Gauckler and C. Lubich, Nonlinear Schrödinger equations and their spectral semi-discretizations over long times, *Found. Comput. Math.* **10**(2), 141–169 (2010).

[50] E. Faou, B. Grébert, and E. Paturel, Birkhoff normal form for splitting methods applied to semilinear Hamiltonian PDEs. I. Finite-dimensional discretization, *Numer. Math.* **114**(3), 429–458 (2010).

[51] E. Faou, B. Grébert, and E. Paturel, Birkhoff normal form for splitting methods applied to semilinear Hamiltonian PDEs. II. Abstract splitting, *Numer. Math.* **114**(3), 459–490 (2010).
[52] E. Faou and B. Grébert, Hamiltonian interpolation of splitting approximations for nonlinear PDEs, *Found. Comput. Math.* **11**(4), 381–415 (2011).
[53] Y. Wang and X. Zhao, A symmetric low-regularity integrator for nonlinear Klein-Gordon equation, *Math. Comp.* **91**(337), 2215–2245 (2022).
[54] G. Maierhofer and K. Schratz, Bridging the gap: symplecticity and low regularity in Runge-Kutta resonance-based schemes, *arXiv:2205.05024*, (2023).
[55] V. Banica, G. Maierhofer, and K. Schratz, Numerical integration of Schrödinger maps via the Hasimoto transform, *SIAM J. Numer. Anal.* **62** (1), 322–352 (2024).
[56] W. Bao and C. Wang, An explicit and symmetric exponential wave integrator for the nonlinear Schrödinger equation with low regularity potential and nonlinearity, *SIAM J. Numer. Anal.* **62**(4), 1901–1928 (2024).
[57] R. Carles and C. Su, Scattering and uniform in time error estimates for splitting method in NLS, *Found. Comput. Math.* **24**(2), 683–722 (2024).
[58] Y. Alama Bronsard, X. Chen, and M. Dolbeault, Spectrally accurate fully discrete schemes for some nonlocal and nonlinear integrable PDEs via explicit formulas, *arXiv:2412.13480*, (2024).
[59] Y. Feng, G. Maierhofer, and K. Schratz, Long-time error bounds of low-regularity integrators for nonlinear Schrödinger equations, *Math. Comp.* **93** (348), 1569–1598 (2024).
[60] Y. Feng, G. Maierhofer, and C. Wang, Explicit symmetric low-regularity integrators for the nonlinear Schrödinger equation, *arXiv:2411.07720* (2024).

https://doi.org/10.1142/9789819816620_0002

Chapter 2

Error Estimates of Numerical Methods for Long-time Dynamics of Nonlinear Klein-Gordon Equation

Weizhu Bao*, Yongyong Cai† and Yue Feng‡

**Department of Mathematics, National University of Singapore, Singapore 119076, Singapore*
matbaowz@nus.edu.sg

†Laboratory of Mathematics and Complex Systems and School of Mathematical Sciences, Beijing Normal University, Beijing 100875, P. R. China
yongyong.cai@bnu.edu.cn

‡School of Mathematics and Statistics, Xi'an Jiaotong University, Xi'an 710049, P. R. China
yue.feng@xjtu.edu.cn

This chapter reviews some recent progress for the study on (uniform) error bounds of numerical methods for the long-time dynamics of the nonlinear Klein-Gordon equation (NKGE) with weak nonlinearity, where the nonlinearity strength is characterized by ε^2 with $\varepsilon \in (0,1]$ a dimensionless parameter up to the time at $O(\varepsilon^{-2})$. Various spatial and temporal discretizations are applied to solve the NKGE including finite difference method and spectral method in space, and finite difference method, exponential integrator and time-splitting method in time. Rigorous error estimates of these numerical methods up to the time at $O(\varepsilon^{-2})$ are presented. Particular attentions are paid to the error bounds of these numerical methods in the long-time regime and how the error bounds depend explicitly on the mesh size, time step as well as the small parameter ε, which indicates how to choose the mesh size and time step in the long-time numerical simulations. Especially, a new numerical analysis technique named as Regularity Compensation Oscillation (RCO) is introduced to establish the improved uniform error bounds of the time-splitting method for the long-time dynamics of the NKGE. In each part, numerical examples are reported to confirm the error estimates and to demonstrate that they are sharp. In addition, comparisons of different numerical methods are concluded and these results are extended to an equivalent oscillatory NKGE and other dispersive PDEs.

Contents

1. The nonlinear Klein-Gordon equation

The nonlinear Klein-Gordon equation (NKGE) is a fundamental equation in relativistic quantum mechanics and quantum field theory to describe the motion of spinless particles [1, 2]. It is the relativistic and nonlinear version of the Schrödinger equation and widely used to model nonlinear phenomena in many fields of science and engineering. It plays a fundamental role in quantum electrodynamics, particle and plasma physics within the framework of quantum mechanics and Einstein's special relativity. In particular, the NKGE with power-type nonlinearity has attracted much attention in studying the relativistic Bose gas, the dynamics of Copper pairs in superconductors and so on [3, 4].

In this chapter, we consider the following NKGE [5–8]

$$\begin{cases} \partial_{tt}u(\mathbf{x},t) - \Delta u(\mathbf{x},t) + u(\mathbf{x},t) + \varepsilon^2 u^3(\mathbf{x},t) = 0, & \mathbf{x} \in \Omega, \quad t > 0, \\ u(\mathbf{x},0) = \phi(\mathbf{x}), \quad \partial_t u(\mathbf{x},0) = \gamma(\mathbf{x}), & \mathbf{x} \in \Omega, \end{cases} \tag{1.1}$$

where t is time, $\mathbf{x}$ is the spatial coordinate, Δ is the Laplace operator, $u(\mathbf{x},t)$ is a real-valued scalar field, $\varepsilon \in (0,1]$ is a dimensionless parameter used to characterize the nonlinearity strength and $\Omega = \prod_{i=1}^{d}(a_i, b_i) \subset \mathbb{R}^d$ ($d = 1, 2, 3$) is a bounded domain equipped with periodic boundary conditions. The initial data $\phi(\mathbf{x})$ and $\gamma(\mathbf{x})$ are two given real-valued functions independent of the parameter ε. The NKGE (1.1) is time symmetric or time reversible. In addition, if $u(\cdot, t) \in H^1(\Omega)$ and $\partial_t u(\cdot, t) \in L^2(\Omega)$, the NKGE (1.1) conserves the energy as [9, 10]

$$\begin{aligned} E(t) &:= E(u(\cdot, t)) \\ &= \int_\Omega \left[|\partial_t u(\mathbf{x}, t)|^2 + |\nabla u(\mathbf{x}, t)|^2 + |u(\mathbf{x}, t)|^2 + \frac{\varepsilon^2}{2} u^4(\mathbf{x}, t) \right] d\mathbf{x} \\ &\equiv \int_\Omega \left[|\gamma(\mathbf{x})|^2 + |\nabla \phi(\mathbf{x})|^2 + |\phi(\mathbf{x})|^2 + \frac{\varepsilon^2}{2} \phi^4(\mathbf{x}, t) \right] d\mathbf{x} \\ &= E(0) = O(1), \quad t \geq 0. \end{aligned} \tag{1.2}$$

When $0 < \varepsilon \ll 1$, by introducing a new variable $w(\mathbf{x}, t) = \varepsilon u(\mathbf{x}, t)$, the NKGE (1.1) with $O(\varepsilon^2)$ nonlinearity and $O(1)$ initial data can be reformulated into the following NKGE with $O(\varepsilon)$ initial data and $O(1)$ nonlinearity:

$$\begin{cases} \partial_{tt} w(\mathbf{x}, t) - \Delta w(\mathbf{x}, t) + w(\mathbf{x}, t) + w^3(\mathbf{x}, t) = 0, & \mathbf{x} \in \Omega, \quad t > 0, \\ w(\mathbf{x}, 0) = \varepsilon \phi(\mathbf{x}), \quad \partial_t w(\mathbf{x}, 0) = \varepsilon \gamma(\mathbf{x}), & \mathbf{x} \in \Omega. \end{cases} \tag{1.3}$$

In this equation, the amplitude of the solution $w(\mathbf{x}, t)$ is at $O(\varepsilon)$ due to the $O(\varepsilon)$ initial data. The wavelength of the solution in space and time is at $O(1)$ and the wave velocity is also at $O(1)$. Similarly, the NKGE (1.3) is also time symmetric or time reversible and conserves the energy as

$$\begin{aligned} \widetilde{E}(t) &:= \widetilde{E}(w(\cdot, t)) \\ &= \int_\Omega \left[|\partial_t w(\mathbf{x}, t)|^2 + |\nabla w(\mathbf{x}, t)|^2 + |w(\mathbf{x}, t)|^2 + \frac{1}{2} w^4(\mathbf{x}, t) \right] d\mathbf{x} \\ &\equiv \varepsilon^2 \int_\Omega \left[|\gamma(\mathbf{x})|^2 + |\nabla \phi(\mathbf{x})|^2 + |\phi(\mathbf{x})|^2 + \frac{\varepsilon^2}{2} \phi^4(\mathbf{x}, t) \right] d\mathbf{x} \\ &= \varepsilon^2 E(0) = O(\varepsilon^2), \quad t \geq 0. \end{aligned} \tag{1.4}$$

In fact, the long-time dynamics of the NKGE (1.3) with $O(\varepsilon)$ initial data and $O(1)$ nonlinearity is equivalent to that of the NKGE (1.1) with $O(\varepsilon^2)$ nonlinearity and $O(1)$ initial data.

There are two dynamical problems related to the time evolution of the NKGE (1.1) or (1.3): (i) when $\varepsilon = \varepsilon_0$ (e.g., $\varepsilon = 1$) fixed, i.e., in the standard nonlinearity strength regime, to study the finite time dynamics of the NKGE (1.1) or (1.3) for $t \in [0, T]$ with $T = O(1)$; and (ii) when $0 < \varepsilon \ll 1$, i.e., in the weak nonlinearity strength regime, to study the long-time dynamics of the NKGE (1.1) or (1.3) for $t \in [0, T_\varepsilon]$ with $T_\varepsilon = O(\varepsilon^{-2})$. In the past several decades, extensive analytical and numerical studies have been done in the literature for the finite time dynamics of the NKGE (1.1) with $\varepsilon = 1$, i.e., in the standard nonlinearity strength regime. Along the analytical front, the existence of global classical solutions, almost periodic solutions and asymptotic behavior of the solution have been widely investigated [6, 11–13]. For the numerical perspective, different numerical methods have been presented and analyzed, such as finite difference time domain methods, exponential integrator Fourier pseudospectral method and time-splitting Fourier pseudospectral method [9, 14–16]. However, the error estimates of these methods are normally valid up to the time at $O(1)$ instead of the long-time regime.

Recently, the long-time dynamics of the NKGE (1.1) in the weak nonlinearity regime or the NKGE (1.3) with small initial data has attracted much attention. According to the analytical results, the life-span of a smooth solution to the NKGE (1.1) or (1.3) is at least up to the time at $O(\varepsilon^{-2})$ [7, 17–20]. For the long-time dynamics, near-conservation of energy, momentum and harmonic actions has been established for the semi-discretization and full-discretization of the NKGE (1.3) with small initial data via the technique of modulated Fourier expansions [10, 21, 22].

In this chapter, we introduce different numerical methods to approximate the NKGE (1.1) and establish the long-time error bounds, especially how the error bounds depend explicitly on the mesh size and time step as well as the parameter ε. We compare the long-time performance of different discretizations in space and time. For simplicity, we introduce the numerical schemes and corresponding error estimates for one dimension (1D) and generalizations to higher dimensions are straightforward. In 1D, the NKGE (1.1) with periodic boundary conditions on the domain $\Omega = (a, b)$ reduces to

$$\begin{cases} \partial_{tt}u(x,t) - \partial_{xx}u(x,t) + u(x,t) + \varepsilon^2 u^3(x,t) = 0, & x \in \Omega,\ t > 0, \\ u(x,0) = \phi(x),\ \partial_t u(x,0) = \gamma(x), & x \in \overline{\Omega} = [a,b], \end{cases} \tag{1.5}$$

with boundary conditions as $u(a,t) = u(b,t)$, $\partial_x u(a,t) = \partial_x u(b,t)$ for $t \geq 0$.

The rest of this chapter is organized as follows. In Section 2, the explicit/semi-implicit/implicit and conservative/nonconservative finite difference time domain methods are presented with error estimates carried out. In Section 3, the exponential integrator Fourier pseudospectral method is shown with uniform error bound in the long-time regime established.

In Section 4, the time-splitting Fourier pseudospectral method is applied to solve the NKGE and the Regularity Compensation Oscillation (RCO) technique is introduced to establish the improved uniform error bounds in the long-time regime. In Section 5, comparisons of these methods are shown and corresponding error estimates are extended to an equivalent oscillatory NKGE and other dispersive PDEs. Finally, some conclusions are drawn in Section 6. Throughout this chapter, the notation $A \lesssim B$ is used to represent that there exists a generic constant $C > 0$ independent of the mesh size h, time step τ and the parameter ε such that $|A| \leq CB$.

2. Finite difference time domain methods

In this section, we present the finite difference time domain (FDTD) methods to discretize the NKGE (1.5) and carry out the long-time error estimates [23].

2.1. *Numerical methods*

Choose a time step $\tau := \Delta t > 0$ and spatial mesh size $h = (b-a)/M$ with M being a positive integer. Define the index sets

$$\mathcal{T}_M = \{j \mid j = 1, 2, \ldots, M-1\}, \quad \mathcal{T}_M^0 = \{j \mid j = 0, 1, 2, \ldots, M\}.$$

We denote the grid points and time steps as

$$x_j := a + jh, \quad j \in \mathcal{T}_M^0; \quad t_n := n\tau, \quad n = 0, 1, 2, \ldots. \tag{2.1}$$

Introduce the space

$$X_M = \left\{u = (u_0, u_1, \ldots, u_M)^T \mid u_j \in \mathbb{R},\ j \in \mathcal{T}_M^0,\ u_0 = u_M\right\},$$

and we always use $u_{-1} = u_{M-1}$ and $u_{M+1} = u_1$ if they are involved. The standard discrete ℓ^2, semi H^1 and ℓ^∞ norms over X_M are defined as

$$\|u\|_{\ell^2}^2 = h \sum_{j=0}^{M-1} |u_j|^2, \quad \|\delta_x^+ u\|_{\ell^2}^2 = h \sum_{j=0}^{M-1} |\delta_x^+ u_j|^2, \quad \|u\|_{\ell^\infty} = \max_{j \in \mathcal{T}_M^0} |u_j|.$$

Let u_j^n be the numerical approximations of $u(x_j, t_n)$ for $j \in \mathcal{T}_M^0$, $n \geq 0$ and denote the numerical solution at time $t = t_n$ as $u^n = (u_0^n, u_1^n, \ldots, u_M^n)^T \in X_M$. We introduce the finite difference operators as

$$\delta_t^+ u_j^n = \frac{u_j^{n+1} - u_j^n}{\tau}, \quad \delta_t^- u_j^n = \frac{u_j^n - u_j^{n-1}}{\tau}, \quad \delta_t^2 u_j^n = \frac{u_j^{n+1} - 2u_j^n + u_j^{n-1}}{\tau^2},$$

$$\delta_x^+ u_j^n = \frac{u_{j+1}^n - u_j^n}{h}, \quad \delta_x^- u_j^n = \frac{u_j^n - u_{j-1}^n}{h}, \quad \delta_x^2 u_j^n = \frac{u_{j+1}^n - 2u_j^n + u_{j-1}^n}{h^2}.$$

We consider four frequently used FDTD methods to discretize the NKGE (1.5), i.e., for $j = 0, 1, \ldots, M-1$ and $n = 1, 2, \ldots$.

I. The Crank-Nicolson finite difference (CNFD) method

$$\delta_t^2 u_j^n - \frac{1}{2}\delta_x^2 \left(u_j^{n+1} + u_j^{n-1}\right) + \frac{1}{2}\left(u_j^{n+1} + u_j^{n-1}\right) + \varepsilon^2 G\left(u_j^{n+1}, u_j^{n-1}\right) = 0; \tag{2.2}$$

II. A semi-implicit energy conservative finite difference (SIFD1) method

$$\delta_t^2 u_j^n - \delta_x^2 u_j^n + \frac{1}{2}\left(u_j^{n+1} + u_j^{n-1}\right) + \varepsilon^2 G\left(u_j^{n+1}, u_j^{n-1}\right) = 0; \tag{2.3}$$

III. Another semi-implicit finite difference (SIFD2) method

$$\delta_t^2 u_j^n - \frac{1}{2}\delta_x^2 \left(u_j^{n+1} + u_j^{n-1}\right) + \frac{1}{2}\left(u_j^{n+1} + u_j^{n-1}\right) + \varepsilon^2 \left(u_j^n\right)^3 = 0; \tag{2.4}$$

IV. An explicit finite difference (EXFD) method

$$\delta_t^2 u_j^n - \delta_x^2 u_j^n + u_j^n + \varepsilon^2 \left(u_j^n\right)^3 = 0. \tag{2.5}$$

In the CNFD (2.2) and SIFD1 (2.3), the nonlinear term is defined as

$$G(v, w) = \frac{F(v) - F(w)}{v - w}, \quad \forall\, v, w \in \mathbb{R}, \quad F(v) = \int_0^v s^3 ds = \frac{v^4}{4}, \quad v \in \mathbb{R}.$$

The initial and boundary conditions in (1.5) are discretized as

$$u_0^{n+1} = u_M^{n+1}, \quad u_{-1}^{n+1} = u_{M-1}^{n+1}, \quad n \geq 0; \quad u_j^0 = \phi(x_j), \quad j \in \mathcal{T}_M^0, \tag{2.6}$$

where the initial velocity $\gamma(x)$ is employed to update the first step u^1 by the Taylor expansion and NKGE (1.5) as

$$u_j^1 = \phi(x_j) + \tau\gamma(x_j) + \frac{\tau^2}{2}\left[\delta_x^2 \phi(x_j) - \phi(x_j) - \varepsilon^2 \left(\phi(x_j)\right)^3\right], \quad j \in \mathcal{T}_M^0. \tag{2.7}$$

It is easy to check that the above FDTD methods are all time symmetric or time reversible, i.e., they are unchanged if interchanging $n+1 \leftrightarrow n-1$ and $\tau \leftrightarrow -\tau$. In addition, the EXFD (2.5) might be the simplest and the most efficient discretization for the NKGE (1.5) with the computational cost per time step at $O(M)$, while the others are implicit schemes. Nevertheless, the CNFD (2.2) and SIFD1 (2.3) can be solved via either a direct solver or an iterative solver with the computational cost per time step depending on the solver, which is usually larger than $O(M)$, especially in two dimensions (2D) and three dimensions (3D). Meanwhile, the solution of the SIFD2 (2.4) can be explicitly updated in the Fourier space with $O(M \ln M)$ computational cost per time step, and such approach is valid in higher dimensions.

According to the von Neumann stability analysis, we have following stability conditions of the above FDTD methods for the NKGE (1.5) [23].

Lemma 1. *Let $T > 0$ be a fixed constant, and denote*

$$\sigma_{\max} := \max_{0 \le n \le \frac{T/\varepsilon^2}{\tau}} \|u^n\|_{\ell^\infty}^2. \tag{2.8}$$

For the above FDTD methods applied to solve the NKGE (1.5) up to the time $t = T/\varepsilon^2$, we have:

(i) The CNFD (2.2) is unconditionally stable for any $h > 0, \tau > 0$ and $0 < \varepsilon \le 1$.

(ii) When $h \ge 2$, the SIFD1 (2.3) is unconditionally stable for any $\tau > 0$; and when $0 < h < 2$, this scheme is conditionally stable under the stability condition

$$0 < \tau < \frac{2h}{\sqrt{4 - h^2}}, \quad 0 < \varepsilon \le 1. \tag{2.9}$$

(iii) When $\sigma_{\max} \le \varepsilon^{-2}$, the SIFD2 (2.4) is unconditionally stable for any $h > 0$ and $\tau > 0$; and when $\sigma_{\max} > \varepsilon^{-2}$, this scheme is conditionally stable under the stability condition

$$0 < \tau < \frac{2}{\sqrt{\varepsilon^2 \sigma_{\max} - 1}}, \quad h > 0, \quad 0 < \varepsilon \le 1. \tag{2.10}$$

(iv) The EXFD (2.5) is conditionally stable under the stability condition

$$0 < \tau < \frac{2h}{\sqrt{4 + h^2(1 + \varepsilon^2 \sigma_{\max})}}, \quad h > 0, \quad 0 < \varepsilon \le 1. \tag{2.11}$$

2.2. *Long-time error estimates*

We establish the error bounds of the FDTD methods for the NKGE (1.5) up to the time at $O(1/\varepsilon^2)$. We assume that the exact solution $u(x,t)$ of the NKGE (1.5) up to the time at $T_\varepsilon = T/\varepsilon^2$ with $T > 0$ fixed satisfies:

$$\text{(A)} \quad \begin{aligned} &u \in C([0,T_\varepsilon]; W^{4,\infty}) \cap C^2([0,T_\varepsilon]; W^{2,\infty}) \\ &\qquad \cap C^3([0,T_\varepsilon]; W^{1,\infty}) \cap C^4([0,T_\varepsilon]; L^\infty), \\ &\left\| \frac{\partial^{r+q}}{\partial t^r \partial x^q} u(x,t) \right\|_{L^\infty} \lesssim 1, \quad 0 \le r \le 4, \quad 0 \le r+q \le 4. \end{aligned}$$

Denote $M_1 = \sup_{\varepsilon \in (0,1]} \|u(x,t)\|_{L^\infty}$ and the grid "error" function $e^n \in X_M (n \ge 0)$ as

$$e_j^n = u(x_j, t_n) - u_j^n, \quad j \in \mathcal{T}_M^0, \quad n = 0, 1, 2, \ldots, \tag{2.12}$$

where $u^n \in X_M$ is the numerical approximation of the NKGE (1.5). We can establish the following error estimates for the NKGE (1.5) in the long-time regime.

Theorem 2. *Under the assumption (A), there exist constants $h_0 > 0$ and $\tau_0 > 0$ sufficiently small and independent of ε, such that, for any $0 < \varepsilon \leq 1$, when $0 < h \leq \varepsilon h_0$ and $0 < \tau \leq \varepsilon \tau_0$, we have the following error estimates for the CNFD (2.2) with (2.6) and (2.7)*

$$\|e^n\|_{\ell^2} + \|\delta_x^+ e^n\|_{\ell^2} \lesssim \frac{h^2}{\varepsilon^2} + \frac{\tau^2}{\varepsilon^2}, \quad \|u^n\|_{\ell^\infty} \leq 1 + M_1, \quad 0 \leq n \leq \frac{T/\varepsilon^2}{\tau}. \tag{2.13}$$

Theorem 3. *Assume $\tau \lesssim h$ and under the assumption (A), there exist constants $h_0 > 0$ and $\tau_0 > 0$ sufficiently small and independent of ε, such that for any $0 < \varepsilon \leq 1$, when $0 < h \leq \varepsilon h_0$, $0 < \tau \leq \varepsilon \tau_0$ and under the stability condition, we have the following error estimates for the SIFD1 (2.3) with (2.6) and (2.7)*

$$\|e^n\|_{\ell^2} + \|\delta_x^+ e^n\|_{\ell^2} \lesssim \frac{h^2}{\varepsilon^2} + \frac{\tau^2}{\varepsilon^2}, \quad \|u^n\|_{\ell^\infty} \leq 1 + M_1, \quad 0 \leq n \leq \frac{T/\varepsilon^2}{\tau}. \tag{2.14}$$

Theorem 4. *Assume $\tau \lesssim h$ and under the assumption (A), there exist constants $h_0 > 0$ and $\tau_0 > 0$ sufficiently small and independent of ε, such that for any $0 < \varepsilon \leq 1$, when $0 < h \leq \varepsilon h_0$, $0 < \tau \leq \varepsilon \tau_0$ and under the stability condition, we have the following error estimates for the SIFD2 (2.4) with (2.6) and (2.7)*

$$\|e^n\|_{\ell^2} + \|\delta_x^+ e^n\|_{\ell^2} \lesssim \frac{h^2}{\varepsilon^2} + \frac{\tau^2}{\varepsilon^2}, \quad \|u^n\|_{\ell^\infty} \leq 1 + M_1, \quad 0 \leq n \leq \frac{T/\varepsilon^2}{\tau}. \tag{2.15}$$

Theorem 5. *Assume $\tau \leq \frac{1}{2}\min\{1, h\}$ and under the assumption (A), there exist constants $h_0 > 0$ and $\tau_0 > 0$ sufficiently small and independent of ε, such that for any $0 < \varepsilon \leq 1$, when $0 < h \leq \varepsilon h_0$ and $0 < \tau \leq \varepsilon \tau_0$ and under the stability condition, we have the following error estimates for the EXFD (2.5) with (2.6) and (2.7)*

$$\|e^n\|_{\ell^2} + \|\delta_x^+ e^n\|_{\ell^2} \lesssim \frac{h^2}{\varepsilon^2} + \frac{\tau^2}{\varepsilon^2}, \quad \|u^n\|_{\ell^\infty} \leq 1 + M_1, \quad 0 \leq n \leq \frac{T/\varepsilon^2}{\tau}. \tag{2.16}$$

Based on above error estimates, these FDTD methods have the same spatial and temporal resolution capacity for the NKGE (1.5) up to the long-time at $O(\varepsilon^{-2})$. In fact, given an accuracy bound $\delta_0 > 0$, the ε-scalability (or meshing strategy) of the FDTD methods should be taken as

$$h = O(\varepsilon\sqrt{\delta_0}) = O(\varepsilon), \quad \tau = O(\varepsilon\sqrt{\delta_0}) = O(\varepsilon), \quad 0 < \varepsilon \le 1.$$

This implies that in order to get "correct" numerical solution up to the time at $O(\varepsilon^{-2})$, one has to take the meshing strategy: $h = O(\varepsilon)$ and $\tau = O(\varepsilon)$. These results are very useful for practical computations on how to select mesh size and time step such that the numerical results are trustable.

Remark 6. In 2D ($d = 2$) and 3D ($d = 3$) cases, the above theorems are still valid under the technical conditions $0 < h \lesssim \varepsilon\sqrt{C_d(h)}$ and $0 < \tau \lesssim \varepsilon\sqrt{C_d(h)}$ where $C_d(h) = 1/|\ln h|$ when $d = 2$, and $C_d(h) = h^{1/2}$ when $d = 3$.

Remark 7. In the previous part, the second-order finite difference is used to discretize the NKGE (1.5) in space. We can also use fourth-order compact finite difference method in space, which leads to the fourth-order compact finite difference (4cFD) methods [24]. By similar error estimates, the 4cFD methods have the following spatial and temporal resolution capacity for the NKGE (1.5) in the long-time regime. In fact, given an accuracy bound $\delta_0 > 0$, the ε-scalability of the 4cFD methods is:

$$h = O(\varepsilon^{1/2}\sqrt{\delta_0}) = O(\varepsilon^{1/2}), \quad \tau = O(\varepsilon\sqrt{\delta_0}) = O(\varepsilon), \quad 0 < \varepsilon \le 1.$$

Compared with the FDTD methods, the 4cFD methods can attain higher order accuracy in space for a given mesh size or improve the spatial resolution capacity, i.e., it needs less grid points while maintaining the same accuracy.

In addition, for the CNFD (2.2) and SIFD1 (2.3), we can show that they conserve the energy in the discretized level [9].

Theorem 8 (Energy conservation). *For $n \ge 0$, the CNFD (2.2) conserves the discrete energy as*

$$\begin{aligned} E_1^n &:= \|\delta_t^+ u^n\|_{\ell^2}^2 + \frac{1}{2}\left(\|\delta_x^+ u^n\|_{\ell^2}^2 + \|\delta_x^+ u^{n+1}\|_{\ell^2}^2\right) + \frac{1}{2}\left(\|u^n\|_{\ell^2}^2 + \|u^{n+1}\|_{\ell^2}^2\right) \\ &\quad + \frac{\varepsilon^2 h}{4}\sum_{j=0}^{M-1}\left[(u_j^n)^4 + (u_j^{n+1})^4\right] \\ &\equiv E_1^0. \end{aligned}$$

Similarly, the SIFD1 (2.3) conserves the discrete energy as

$$\begin{aligned} E_2^n &:= \|\delta_t^+ u^n\|_{\ell^2}^2 + h\sum_{j=0}^{M-1} \left(\delta_x^+ u_j^n\right)\left(\delta_x^+ u_j^{n+1}\right) + \frac{1}{2}\left(\|u^n\|_{\ell^2}^2 + \|u^{n+1}\|_{\ell^2}^2\right) \\ &\quad + \frac{\varepsilon^2 h}{4}\sum_{j=0}^{M-1}\left[(u_j^n)^4 + (u_j^{n+1})^4\right] \\ &\equiv E_2^0, \qquad n \geq 0. \end{aligned}$$

2.3. *Numerical results*

In this subsection, we present some numerical results of the FDTD methods for the NKGE (1.5) up to the time at $O(\varepsilon^{-2})$ to verify our error bounds. We only show numerical results for the CNFD (2.2) and the results for other FDTD methods are quite similar which are omitted for brevity. In the numerical experiments, we take $a = 0$, $b = 2\pi$ and choose the initial data as

$$\phi(x) = \cos(x) + \cos(2x), \qquad \gamma(x) = \sin(x), \qquad 0 \leq x \leq 2\pi. \tag{2.17}$$

The "exact" solution is obtained numerically by the exponential integrator Fourier pseudospectral method with a very fine mesh size and a very small time step, e.g. $h_e = \pi/2^{15}$ and $\tau_e = 10^{-5}$ [25]. Denote $u_{h,\tau}^n$ as the numerical solution at time $t = t_n$ obtained by a numerical method with mesh size h and time step τ. In order to quantify the numerical results, we define the error function as follows:

$$e_{h,\tau}(t_n) = \sqrt{\|u(\cdot,t_n) - u_{h,\tau}^n\|_{\ell^2}^2 + \|\delta_x^+(u(\cdot,t_n) - u_{h,\tau}^n)\|_{\ell^2}^2}. \tag{2.18}$$

We first test the spatial discretization errors at $t = 1/\varepsilon^2$ for different $0 < \varepsilon \leq 1$. In order to do this, we fix the time step as $\tau_e = 10^{-5}$ such that the temporal error can be ignored, and solve the NKGE (1.5) under different mesh size h. Then we check the temporal errors at $t = 1/\varepsilon^2$ for different $0 < \varepsilon \leq 1$ with different time step τ and a fine mesh size $h_e = \pi/2^{11}$ such that the spatial errors can be neglected. Table 1 and Table 2 depict the spatial and temporal errors of the CNFD (2.2) for the NKGE (1.5) with the initial data (2.17) in the long-time regime, respectively.

From Tables 1 and 2 for the CNFD method and additional similar numerical results for other FDTD methods not shown here for brevity, we can draw the following observations:

(i) For any fixed $\varepsilon = \varepsilon_0 > 0$, the FDTD methods are uniformly second-order accurate in both spatial and temporal discretizations (cf. the first rows in Tables 1 and 2), which agree with those results in the literature.

Table 1. Spatial errors of the CNFD (2.2) for the NKGE (1.5).

$e_{h,\tau_e}(t=1/\varepsilon^2)$	$h_0=\pi/16$	$h_0/2$	$h_0/2^2$	$h_0/2^3$	$h_0/2^4$	$h_0/2^5$
$\varepsilon_0=1$	**3.77E-2**	9.65E-3	2.43E-3	6.09E-4	1.52E-4	3.84E-5
order	**-**	1.97	1.99	2.00	2.00	1.98
$\varepsilon_0/2$	3.98E-2	**9.56E-3**	2.39E-3	5.97E-4	1.49E-4	3.81E-5
order	-	**2.06**	2.00	2.00	2.00	1.97
$\varepsilon_0/2^2$	7.17E-1	1.82E-1	**4.55E-2**	1.14E-2	2.85E-3	7.27E-4
order	-	1.98	**2.00**	2.00	2.00	1.97
$\varepsilon_0/2^3$	2.78E+0	6.54E-1	1.58E-1	**3.92E-2**	9.78E-3	2.50E-3
order	-	2.09	2.05	**2.01**	2.00	1.97
$\varepsilon_0/2^4$	3.31E+0	1.78E+0	5.92E-1	1.55E-1	**3.93E-2**	1.01E-2
order	-	0.89	1.59	1.93	**1.98**	1.96

Table 2. Temporal errors of the CNFD (2.2) for the NKGE (1.5).

$e_{h_e,\tau}(t=1/\varepsilon^2)$	$\tau_0=0.05$	$\tau_0/2$	$\tau_0/2^2$	$\tau_0/2^3$	$\tau_0/2^4$	$\tau_0/2^5$
$\varepsilon_0=1$	**3.27E-2**	8.57E-3	2.19E-3	5.53E-4	1.39E-4	3.48E-5
order	**-**	1.93	1.97	1.99	1.99	2.00
$\varepsilon_0/2$	2.56E-2	**6.32E-3**	1.58E-3	3.94E-4	9.86E-5	2.47E-5
order	-	**2.02**	2.00	2.00	2.00	2.00
$\varepsilon_0/2^2$	3.91E-1	9.83E-2	**2.46E-2**	6.16E-3	1.54E-3	3.85E-4
order	-	1.99	**2.00**	2.00	2.00	2.00
$\varepsilon_0/2^3$	1.40E+0	3.32E-1	8.14E-2	**2.03E-2**	5.06E-3	1.26E-3
order	-	2.08	2.03	**2.00**	2.00	2.01
$\varepsilon_0/2^4$	1.81E+0	1.13E+0	3.16E-1	8.07E-2	**2.03E-2**	5.07E-3
order	-	0.68	1.84	1.97	**1.99**	2.00

(ii) In the long-time regime, the second-order convergence in space and time of the FDTD methods can be observed only when $0 < h \lesssim \varepsilon$ and $0 < \tau \lesssim \varepsilon$ (cf. upper triangles above the diagonals (corresponding to $h \sim \varepsilon$ and $\tau \sim \varepsilon$, and being labeled in bold letters) in Tables 1 and 2), which again confirms our error bounds.

In summary, these numerical results confirm the error bounds of the FDTD methods and show that they are sharp.

3. Exponential integrator Fourier pseudospectral method

In this section, we introduce the exponential integrator Fourier pseudospectral (EIFP) method to discretize the NKGE (1.5) and establish the uniform error bounds of the EIFP method in the long-time regime [25].

3.1. *Numerical methods*

Choose a time step $\tau := \Delta t > 0$ and mesh size $h := \Delta x = (b-a)/N$ with N being an even positive integer. Define the index set

$$\mathcal{J}_N = \left\{-\frac{N}{2}, \ldots, \frac{N}{2} - 1\right\},$$

and two spaces $C_{\text{per}}(\Omega) = \{u \in C(\overline{\Omega}) \mid u(a) = u(b)\}$ and

$$Y_N := \text{span}\left\{e^{i\mu_l(x-a)},\ x \in \overline{\Omega},\ l \in \mathcal{J}_N\right\},$$

where $\mu_l = \frac{2\pi l}{b-a}$. For any $u(x) \in C_{\text{per}}(\Omega)$ and a vector $u \in X_N$, let $P_N : L^2(\Omega) \to Y_N$ be the standard L^2-projection operator onto Y_N, $I_N : C_{\text{per}}(\Omega) \to Y_N$ or $I_N : X_N \to Y_N$ be the trigonometric interpolation operator [26], i.e.,

$$P_N u = \sum_{l \in \mathcal{J}_N} \widehat{u}_l e^{i\mu_l(x-a)}, \quad I_N u = \sum_{l \in \mathcal{J}_N} \widetilde{u}_l e^{i\mu_l(x-a)}, \quad x \in \overline{\Omega},$$

where

$$\widehat{u}_l = \frac{1}{b-a}\int_a^b u(x)e^{-i\mu_l(x-a)}dx, \quad \widetilde{u}_l = \frac{1}{N}\sum_{j=0}^{N-1} u_j e^{-i\mu_l(x_j-a)}, \quad l \in \mathcal{J}_N,$$

with u_j interpreted as $u(x_j)$ when involved.

We begin with the Fourier spectral method for discretizing the NKGE (1.5) in space, i.e., to find

$$u_N(x,t) = \sum_{l \in \mathcal{J}_N} \widehat{(u_N)}_l(t) e^{i\mu_l(x-a)} \in Y_N, \quad x \in \overline{\Omega}, \quad t \geq 0, \tag{3.1}$$

such that

$$\partial_{tt}u_N(x,t) - \partial_{xx}u_N(x,t) + u_N(x,t) + \varepsilon^2 P_N f(u_N(x,t)) = 0, \tag{3.2}$$

with $f(v) = v^3$. Plugging (3.1) into (3.2) and noticing the orthogonality of the Fourier basis functions, we find

$$\frac{d^2}{dt^2}\widehat{(u_N)}_l(t) + (1+\mu_l^2)\widehat{(u_N)}_l(t) + \varepsilon^2(\widehat{f(u_N)})_l(t) = 0,\ l \in \mathcal{J}_N,\ t \geq 0. \tag{3.3}$$

For each l, when t is near $t_n (n \geq 0)$, the above ODEs can be rewritten as

$$\frac{d^2}{d\theta^2}\widehat{(u_N)}_l(t_n+\theta) + \zeta_l^2\widehat{(u_N)}_l(t_n+\theta) + \varepsilon^2 \widehat{f}_l^n(\theta) = 0, \quad \theta \in \mathbb{R}, \tag{3.4}$$

where

$$\zeta_l = \sqrt{1+\mu_l^2}, \quad \widehat{f}_l^n(\theta) = (\widehat{f(u_N)})_l(t_n+\theta). \tag{3.5}$$

By using the variation-of-constants formula for the ODEs (3.4), the general solutions of the above second-order ODEs can be written as

$$\begin{aligned}\widehat{(u_N)}_l(t_n+\theta) = {} & c_l^n \cos(\theta\zeta_l) + d_l^n \frac{\sin(\theta\zeta_l)}{\zeta_l} \\ & - \frac{\varepsilon^2}{\zeta_l}\int_0^\theta \widehat{f}_l^n(\omega)\sin(\zeta_l(\theta-\omega))d\omega,\end{aligned} \tag{3.6}$$

where c_l^n and d_l^n are two constants to be determined. When $n = 0$, we consider the solution (3.6) for $\theta \in [0, \tau]$. The initial conditions in the NKGE (1.5) imply

$$\widehat{(u_N)}_l(0) = \widehat{\phi}_l, \quad \frac{d}{dt}\widehat{(u_N)}_l(0) = \widehat{\gamma}_l, \quad l \in \mathcal{J}_N. \tag{3.7}$$

Plugging (3.7) into (3.6) with $n = 0$ and then letting $\theta = \tau$ to determine c_l^0 and d_l^0, we get

$$\widehat{(u_N)}_l(\tau) = \widehat{\phi}_l\cos(\tau\zeta_l) + \widehat{\gamma}_l\frac{\sin(\tau\zeta_l)}{\zeta_l} - \frac{\varepsilon^2}{\zeta_l}\int_0^\tau \widehat{f}_l^0(\omega)\sin(\zeta_l(\tau-\omega))d\omega. \tag{3.8}$$

When $n > 0$, we consider the solution (3.6) for $\theta \in [-\tau, \tau]$ and require the solution to be continuous at $t = t_n$ and $t = t_{n-1} = t_n - \tau$. Plugging $\theta = 0$ and $\theta = -\tau$ into (3.6) and letting $\theta = \tau$ to determine c_l^n and d_l^n, we obtain

$$\begin{aligned}\widehat{(u_N)}_l(t_{n+1}) = {} & -\widehat{(u_N)}_l(t_{n-1}) + 2\cos(\tau\zeta_l)\widehat{(u_N)}_l(t_n) \\ & - \frac{\varepsilon^2}{\zeta_l}\int_0^\tau \left[\widehat{f}_l^n(-\omega) + \widehat{f}_l^n(\omega)\right]\sin(\zeta_l(\tau-\omega))d\omega.\end{aligned} \tag{3.9}$$

In order to design an explicit scheme, we approximate the integrals in (3.8)–(3.9) by the Gautschi-type quadrature [27], i.e., for $l \in \mathcal{J}_N$ and $n \geq 1$,

$$\begin{aligned}\int_0^\tau \widehat{f}_l^0(\omega)\sin(\zeta_l(\tau-\omega))d\omega &\approx \widehat{f}_l^0(0)\int_0^\tau \sin(\zeta_l(\tau-\omega))d\omega \\ &= \frac{\widehat{f}_l^0(0)}{\zeta_l}\left[1-\cos(\tau\zeta_l)\right],\end{aligned}$$

$$\int_0^\tau \left[\widehat{f}_l^n(-\omega) + \widehat{f}_l^n(\omega)\right]\sin(\zeta_l(\tau-\omega))d\omega \approx \frac{2\widehat{f}_l^n(0)}{\zeta_l}\left[1-\cos(\tau\zeta_l)\right].$$

Let $u_N^n(\cdot)$ be the approximations of $u_N(x, t_n)$ for $n \geq 0$. Then the Gautschi-type exponential integrator Fourier spectral (EIFS) method for the NKGE (1.5) reads

$$\begin{aligned} \widehat{(u_N^{n+1})}_l &= -\widehat{(u_N^{n-1})}_l + 2p_l\widehat{(u_N^n)}_l + 2r_l\widehat{(f(u_N^n))}_l, \quad l \in \mathcal{T}_N, \quad n \geq 1, \\ \widehat{(u_N^1)}_l &= p_l\widehat{\phi}_l + q_l\widehat{\gamma}_l + r_l\widehat{(f(\phi))}_l, \quad l \in \mathcal{T}_N, \\ \widehat{(u_N^0)}_l &= \widehat{\phi}_l, \quad l \in \mathcal{T}_N, \end{aligned} \tag{3.10}$$

where the coefficients are

$$p_l = \cos(\tau\zeta_l), \quad q_l = \frac{\sin(\tau\zeta_l)}{\zeta_l}, \quad r_l = \frac{\varepsilon^2(\cos(\tau\zeta_l) - 1)}{\zeta_l^2}. \tag{3.11}$$

However, in practice, it is difficult to compute the Fourier coefficients in (3.10). We simply replace the projections by the interpolations, which refers to the Fourier pseudospectral discretization. Let u_j^n be the approximation of $u(x_j, t_n)$ for $j \in \mathcal{T}_N^0$ and $n \geq 0$ and denote $u^n = (u_0^n, u_1^n, \ldots, u_N^n)^T \in \mathbb{R}^{N+1}$. Choose $u_j^0 = \phi(x_j)$, then for $n \geq 0$, the Gautschi-type exponential integrator Fourier pseudospectral (EIFP) discretization for the NKGE (1.5) is given as

$$u_j^{n+1} = \sum_{l \in \mathcal{T}_N} \widetilde{(u^{n+1})}_l e^{i\mu_l(x_j - a)}, \tag{3.12}$$

with

$$\begin{aligned} \widetilde{(u^{n+1})}_l &= -\widetilde{(u^{n-1})}_l + 2p_l\widetilde{(u^n)}_l + 2r_l\widetilde{(f(u^n))}_l, \quad l \in \mathcal{T}_N, \quad n \geq 1, \\ \widetilde{(u^1)}_l &= p_l\widetilde{\phi}_l + q_l\widetilde{\gamma}_l + r_l\widetilde{(f(\phi))}_l, \quad l \in \mathcal{T}_N, \end{aligned} \tag{3.13}$$

where the coefficients p_l, q_l and r_l are given in (3.11).

The EIFP (3.12)–(3.13) is explicit, time symmetric and easy to extend to 2D and 3D. The memory cost is $O(N)$ and the computational cost per time step is $O(N \ln N)$.

3.2. *Uniform error bounds in the long-time regime*

Let $T > 0$ be a fixed constant, and denote

$$\sigma_{\max} := \max_{0 \leq n \leq \frac{T/\varepsilon^2}{\tau}} \|u^n\|_{\ell^\infty}^2. \tag{3.14}$$

According to the standard von Neumann stability analysis [28], we have the following stability results of the EIFP (3.12)–(3.13) for the NKGE (1.5).

Lemma 9 (Stability). *The EIFP* (3.12)–(3.13) *is conditionally stable under the stability condition*

$$0 < \tau \leq \frac{2h}{\sqrt{\pi^2 + h^2(1 + \varepsilon^2\sigma_{\max})}}, \quad h > 0, \quad 0 < \varepsilon \leq 1. \tag{3.15}$$

We define periodic Sobolev spaces as

$$H^m_{\rm per}(\Omega) := \{u \in H^m(\Omega) : u^{(k)}(a) = u^{(k)}(b),\ k = 0, \dots, m-1\}, \quad m \geq 1.$$

Assume that the exact solution $u(x,t)$ of the NKGE (1.5) up to the time $T_\varepsilon = T/\varepsilon^2$ with $T > 0$ fixed satisfies

$$\text{(B)} \qquad \begin{aligned} &\|u(x,t)\|_{L^\infty([0,T_\varepsilon];L^\infty \cap H^m_{\rm per})} \lesssim 1, \quad m \geq 2, \\ &\|\partial_t u(x,t)\|_{L^\infty([0,T_\varepsilon];W^{1,4})} \lesssim 1, \quad \|\partial_{tt} u(x,t)\|_{L^\infty([0,T_\varepsilon];H^1)} \lesssim 1. \end{aligned}$$

Under the assumption (B), denoting

$$M_1 := \max_{\varepsilon \in (0,1]} \{\|u(x,t)\|_{L^\infty([0,T_\varepsilon];L^\infty)} + \|\partial_t u(x,t)\|_{L^\infty([0,T_\varepsilon];L^\infty)}\} \lesssim 1,$$

and assuming

$$\tau \leq \min\left\{\frac{1}{8}, \frac{\pi h}{3\sqrt{\pi^2 + h^2}}\right\}, \tag{3.16}$$

we can establish the following error bounds of the EIFP (3.12)–(3.13).

Theorem 10. *Let u^n be the approximation obtained from the EIFP* (3.12)–(3.13), *under the stability condition* (3.15) *and the assumptions (B) and* (3.16), *there exist constants $h_0 > 0$ and $\tau_0 > 0$ sufficiently small and independent of ε, such that for any $0 < \varepsilon \leq 1$, when $0 < h \leq h_0$, $0 < \tau \leq \tau_0$, we have*

$$\begin{aligned} &\|u(\cdot, t_n) - I_N u^n\|_{H^1} \lesssim h^{m-1} + \tau^2, \\ &\|I_N u^n\|_{H^1} \leq 1 + M_1, \quad 0 \leq n \leq \frac{T/\varepsilon^2}{\tau}. \end{aligned} \tag{3.17}$$

These results are very useful in practical computations on how to select mesh size and time step such that the numerical results are trustable. The error bounds indicate that the ε-scalability of the EIFP (3.12)–(3.13) up to the time at $O(\varepsilon^{-2})$ is uniform in terms of ε and should be taken as:

$$h = O(1), \quad \tau = O(1), \quad 0 < \varepsilon \leq 1.$$

Remark 11. We can also combine the Fourier pseudospectral method in space and finite difference method in time to solve the NKGE (1.5) in the long-time regime, which is finite difference Fourier pseudospectral (FDFP) method. By similar error estimates, the ε-scalability of the FDFP method up to the time at $O(\varepsilon^{-2})$ is

$$h = O(1), \quad \tau = O(\varepsilon), \quad 0 < \varepsilon \leq 1.$$

3.3. *Numerical results*

In this subsection, we present numerical results of the EIFP (3.12)–(3.13) for the NKGE (1.5) to support our error estimates. In the numerical experiments, we choose the initial data

$$\phi(x) = \frac{1}{2+\cos^2(x)}, \quad \gamma(x) = \sin(x), \quad x \in [0, 2\pi]. \tag{3.18}$$

The "exact" solution $u(x,t)$ is computed by the time-splitting Fourier pseudospectral method with a very fine mesh size $h_e = \pi/32$ and a very small time step $\tau_e = 5 \times 10^{-4}$ [29]. Denote $u^n_{h,\tau}$ as the numerical solution at $t = t_n$ by the EIFP (3.12)–(3.13) with mesh size h and time step τ. The errors are denoted as

$$e(t_n) = \left\|u(x,t_n) - I_M u^n_{h,\tau}\right\|_{H^1}.$$

The errors are displayed at $t = 1/\varepsilon^2$ with different ε. Table 3 and Table 4 depict the spatial and temporal errors of the EIFP (3.12)–(3.13), respectively.

Table 3. Spatial errors of the EIFP (3.12)–(3.13) for the NKGE (1.5) with (3.18) for different ε.

$e(1/\varepsilon^2)$	$h_0 = \pi/2$	$h_0/2$	$h_0/2^2$	$h_0/2^3$
$\varepsilon_0 = 1$	4.05E-2	8.80E-3	1.53E-4	7.19E-8
$\varepsilon_0/2$	4.04E-2	8.46E-3	1.40E-4	9.30E-8
$\varepsilon_0/2^2$	6.12E-2	4.18E-3	1.57E-5	6.90E-8
$\varepsilon_0/2^3$	1.01E-1	3.25E-3	1.45E-4	1.35E-7
$\varepsilon_0/2^4$	6.05E-2	1.31E-3	1.34E-4	4.16E-7

From Tables 3 and 4 and additional similar numerical results not shown here for brevity, we can draw the following observations on the EIFP (3.12)–(3.13) for the NKGE (1.5) up to the time at $O(\varepsilon^{-2})$:

(i) In space, the EIFP (3.12)–(3.13) is uniformly spectrally accurate for any $0 < \varepsilon \leq 1$ (cf. each row in Table 3) and the spatial errors are almost independent of ε (cf. each column in Table 3).

(ii) In time, for any fixed $\varepsilon = \varepsilon_0 > 0$, the EIFP (3.12)–(3.13) is second-order accurate (cf. the first rows in Table 4), which agrees with those results in the literature. In addition, Table 4 illustrates that the temporal errors

Table 4. Temporal errors of the EIFP (3.12)–(3.13) for the NKGE (1.5) with (3.18) for different ε.

$e(1/\varepsilon^2)$	$\tau_0 = 0.2$	$\tau_0/2$	$\tau_0/2^2$	$\tau_0/2^3$	$\tau_0/2^4$	$\tau_0/2^5$
$\varepsilon_0 = 1$	4.59E-2	1.13E-2	2.82E-3	7.04E-4	1.76E-4	4.37E-5
order	-	2.02	2.00	2.00	2.00	2.01
$\varepsilon_0/2$	3.17E-2	7.83E-3	1.95E-3	4.88E-4	1.22E-4	3.04E-5
order	-	2.02	2.01	2.00	2.00	2.00
$\varepsilon_0/2^2$	2.51E-2	6.23E-3	1.55E-3	3.88E-4	9.70E-5	2.42E-5
order	-	2.01	2.01	2.00	2.00	2.00
$\varepsilon_0/2^3$	3.28E-2	8.14E-3	2.03E-3	5.08E-4	1.27E-4	3.17E-5
order	-	2.01	2.00	2.00	2.00	2.00
$\varepsilon_0/2^4$	2.50E-2	6.23E-3	1.56E-3	3.89E-4	9.72E-5	2.43E-5
order	-	2.00	2.00	2.00	2.00	2.00
$\varepsilon_0/2^5$	2.88E-2	7.17E-3	1.79E-3	4.47E-4	1.12E-4	2.79E-5
order	-	2.01	2.00	2.00	2.00	2.01

of the EIFP (3.12)–(3.13) behave like $O(\tau^2)$ for the long-time dynamics up to the time at $O(\varepsilon^{-2})$ (cf. each row and column in Table 4).

(iii) Our numerical results confirm the uniform error bounds in Theorem 10 and demonstrate that they are sharp.

4. Time-splitting Fourier pseudospectral method

In this section, we present the time-splitting Fourier pseudospectral (TSFP) method to discretize the NKGE (1.5) and establish the improved uniform error bounds up to the time at $O(\varepsilon^{-2})$ [29, 30].

4.1. *Numerical methods*

Define the operator $\langle\nabla\rangle = \sqrt{1-\Delta}$ through its action in the Fourier space by:

$$\langle\nabla\rangle u(x) = \sum_{l\in\mathbb{Z}} \sqrt{1+\mu_l^2}\,\widehat{u}_l e^{i\mu_l(x-a)}, \quad \text{for} \quad u(x) = \sum_{l\in\mathbb{Z}} \widehat{u}_l e^{i\mu_l(x-a)}, \quad x \in \overline{\Omega},$$

and the inverse operator $\langle\nabla\rangle^{-1}$ as $\langle\nabla\rangle^{-1}u(x) = \sum_{l\in\mathbb{Z}} \frac{\widehat{u}_l}{\sqrt{1+\mu_l^2}} e^{i\mu_l(x-a)}$, which leads to

$$\left\|\langle\nabla\rangle^{-1}u\right\|_{H^s} = \|u\|_{H^{s-1}} \le \|u\|_{H^s}.$$

Introduce new variables $v(x,t) = \partial_t u(x,t)$ and

$$\psi(x,t) = u(x,t) - i\langle\nabla\rangle^{-1}v(x,t), \quad x \in \overline{\Omega}, \quad t \geq 0, \tag{4.1}$$

then the NKGE (1.5) can be reformulated into the following relativistic nonlinear Schrödinger equation (NLSE) for $\psi := \psi(t) = \psi(x,t)$ (spatial variable x may be omitted for brevity) as

$$\begin{cases} i\partial_t\psi + \langle\nabla\rangle\psi + \dfrac{\varepsilon^2}{8}\langle\nabla\rangle^{-1}\left(\psi + \overline{\psi}\right)^3 = 0, \ x \in \Omega, \ t > 0, \\ \psi(a,t) = \psi(b,t), \quad \partial_x\psi(a,t) = \partial_x\psi(b,t), \quad t \geq 0, \\ \psi(x,0) = \psi_0(x) := \phi(x) - i\langle\nabla\rangle^{-1}\gamma(x), \quad x \in \overline{\Omega}, \end{cases} \tag{4.2}$$

where $\overline{\psi}$ denotes the complex conjugate of ψ. According to (4.1), the solution of the NKGE (1.5) can be recovered by

$$\begin{aligned} u(x,t) &= \frac{1}{2}\left(\psi(x,t) + \overline{\psi}(x,t)\right), \\ v(x,t) &= \frac{i}{2}\langle\nabla\rangle\left(\psi(x,t) - \overline{\psi}(x,t)\right). \end{aligned} \tag{4.3}$$

We remark here that the NKGE (1.5) can also be reformulated as the following first-order (in time) PDEs:

$$\begin{cases} \partial_t u(x,t) - v(x,t) = 0, \quad x \in \Omega, \quad t > 0, \\ \partial_t v(x,t) - \partial_{xx}u(x,t) + u(x,t) + \varepsilon^2 u^3(x,t) = 0, \ x \in \Omega, \ t > 0, \\ u(a,t) = u(b,t), \quad \partial_x u(a,t) = \partial_x u(b,t), \quad t \geq 0, \\ u(x,0) = \phi(x), \quad v(x,0) = \gamma(x), \qquad x \in \overline{\Omega}. \end{cases} \tag{4.4}$$

In fact, the relativistic NLSE (4.2) can be decomposed into the following two subproblems via the time-splitting technique [31, 32]

$$\begin{cases} i\partial_t\psi(x,t) + \langle\nabla\rangle\psi(x,t) = 0, \quad x \in \Omega, \quad t > 0, \\ \psi(a,t) = \psi(b,t), \quad \partial_x\psi(a,t) = \partial_x\psi(b,t), \quad t \geq 0, \\ \psi(x,0) = \psi_0(x), \quad x \in \overline{\Omega}, \end{cases} \tag{4.5}$$

and

$$\begin{cases} i\partial_t\psi(x,t) + \varepsilon^2\langle\nabla\rangle^{-1} f\Big(\dfrac{1}{2}(\psi + \overline{\psi})\Big)(x,t) = 0, \quad x \in \Omega, \quad t > 0, \\ \psi(x,0) = \psi_0(x), \quad x \in \overline{\Omega}. \end{cases} \tag{4.6}$$

The linear equation (4.5) can be solved exactly in phase space and the associated evolution operator is given by

$$\psi(\cdot,t) = \varphi_T^t(\psi_0) := e^{it\langle\nabla\rangle}\psi_0, \quad t \geq 0, \tag{4.7}$$

which satisfies the isometry relation

$$\|\varphi_T^t(v_0)\|_{H^s} = \|v_0\|_{H^s}, \quad s \geq 0, \quad t \in \mathbb{R}.$$

Recalling that the nonlinear part of (4.6) is real, this implies that $\partial_t \left(\psi + \overline{\psi}\right)(x,t) = 0$ for any fixed $x \in \overline{\Omega}$. Thus $\psi + \overline{\psi}$ is invariant in time, i.e.,

$$\left(\psi + \overline{\psi}\right)(x,t) \equiv \left(\psi + \overline{\psi}\right)(x,0) = \psi_0(x) + \overline{\psi_0}(x), \quad t \geq 0, \quad x \in \overline{\Omega}. \quad (4.8)$$

Plugging (4.8) into (4.6), we get

$$\begin{cases} i\partial_t \psi(x,t) + \varepsilon^2 \langle \nabla \rangle^{-1} f\Big(\dfrac{1}{2}(\psi_0 + \overline{\psi_0})\Big)(x) = 0, & x \in \overline{\Omega}, \quad t > 0, \\ \psi(x,0) = \psi_0(x), \quad x \in \overline{\Omega}. \end{cases} \quad (4.9)$$

Thus (4.9) or (4.6) can be integrated exactly in time as:

$$\psi(x,t) = \varphi_V^t(\psi_0) := \psi_0(x) + \varepsilon^2 t\, F(\psi_0(x)), \quad t \geq 0, \quad (4.10)$$

where the operator F is defined by

$$F(\phi) = i\langle \nabla \rangle^{-1} B(\phi), \qquad B(\phi) = f\Big(\frac{1}{2}(\phi + \overline{\phi})\Big). \quad (4.11)$$

Let $\tau > 0$ be the time step size and $t_n = n\tau$ $(n = 0, 1, \ldots)$ as the time steps. Denote $\psi^{[n]} := \psi^{[n]}(x)$ as the approximation of $\psi(x, t_n)$, then the second-order discrete-in-time splitting method via the Strang splitting for the relativistic NLSE (4.2) can be written as

$$\begin{aligned} \psi^{[n+1]} &= \mathcal{S}_\tau(\psi^{[n]}) := \varphi_T^{\frac{\tau}{2}} \circ \varphi_V^{\tau} \circ \varphi_T^{\frac{\tau}{2}}(\psi^{[n]}) \\ &= e^{i\tau\langle \nabla \rangle}\psi^{[n]} + \varepsilon^2 \tau e^{i\frac{\tau\langle \nabla \rangle}{2}} F\left(e^{i\frac{\tau\langle \nabla \rangle}{2}}\psi^{[n]}\right), \end{aligned} \quad (4.12)$$

with $\psi^{[0]} = \psi_0 = \phi - i\langle \nabla \rangle^{-1}\gamma$. Noticing (4.3), the semi-discretization of the NKGE (1.5) is given by

$$u^{[n]} = \frac{1}{2}\left(\psi^{[n]} + \overline{\psi^{[n]}}\right), \; v^{[n]} = \frac{i}{2}\langle \nabla \rangle \left(\psi^{[n]} - \overline{\psi^{[n]}}\right), \; n = 0, 1, \ldots, \quad (4.13)$$

where $u^{[n]} := u^{[n]}(x)$ and $v^{[n]} := v^{[n]}(x)$ are the approximations of $u(x, t_n)$ and $\partial_t u(x, t_n)$, respectively.

We remark here that another way to discretize the NKGE (1.5) by a time-splitting method, which is exactly the same discretization as the one presented above, is to discretize the NKGE (4.4) by a time-splitting method. In fact, the NKGE (4.4) can be decomposed into the following two subproblems via the time-splitting technique

$$\begin{cases} \partial_t u(x,t) - v(x,t) = 0, \\ \partial_t v(x,t) - \partial_{xx} u(x,t) + u(x,t) = 0, \quad x \in \Omega, \quad t > 0, \\ u(a,t) = u(b,t), \quad \partial_x u(a,t) = \partial_x u(b,t), \quad t \geq 0, \\ u(x,0) = \phi(x), \quad v(x,0) = \gamma(x), \quad x \in \overline{\Omega}, \end{cases} \quad (4.14)$$

and

$$\begin{cases} \partial_t u(x,t) = 0, \\ \partial_t v(x,t) + \varepsilon^2 u^3(x,t) = 0, \quad x \in \overline{\Omega}, \quad t > 0, \\ u(x,0) = \phi(x), \quad v(x,0) = \gamma(x), \quad x \in \overline{\Omega}. \end{cases} \tag{4.15}$$

Similarly, the linear problem (4.14) can be solved exactly in phase space and the associated evolution operator is given by

$$\begin{pmatrix} u(\cdot,t) \\ v(\cdot,t) \end{pmatrix} = \chi_T^t \begin{pmatrix} \phi \\ \gamma \end{pmatrix} := \begin{pmatrix} \cos(t\langle\nabla\rangle)\phi + \langle\nabla\rangle^{-1}\sin(t\langle\nabla\rangle)\gamma \\ -\langle\nabla\rangle\sin(t\langle\nabla\rangle)\phi + \cos(t\langle\nabla\rangle)\gamma \end{pmatrix}, \quad t \geq 0. \tag{4.16}$$

From (4.15), we obtain immediately that $u(x,t)$ is invariant in time for any fixed $x \in \overline{\Omega}$, i.e.,

$$u(x,t) \equiv u(x,0) = \phi(x), \quad x \in \overline{\Omega}. \tag{4.17}$$

Plugging (4.17) into (4.15), we get

$$\begin{cases} \partial_t u(x,t) = 0, \\ \partial_t v(x,t) + \varepsilon^2 u^3(x,0) = 0, \quad x \in \overline{\Omega}, \quad t > 0, \\ u(x,0) = \phi(x), \quad v(x,0) = \gamma(x), \quad t \geq 0, \quad x \in \overline{\Omega}. \end{cases} \tag{4.18}$$

Thus (4.18) (and (4.15)) can be integrated exactly in time as

$$\begin{pmatrix} u(\cdot,t) \\ v(\cdot,t) \end{pmatrix} = \chi_V^t \begin{pmatrix} \phi \\ \gamma \end{pmatrix} := \begin{pmatrix} \phi \\ \gamma - \varepsilon^2 t\phi^3 \end{pmatrix}, \quad t \geq 0. \tag{4.19}$$

Let $u^{[n]} := u^{[n]}(x)$ and $v^{[n]} := v^{[n]}(x)$ be the approximations of $u(x,t_n)$ and $v(x,t) = \partial_t u(x,t_n)$ $(n = 0,1,2,\ldots)$, respectively, which are the solutions of the NKGE (4.4) (and (1.5)). Then a second-order semi-discretization of the NKGE (4.4) (and (1.5)) via the second-order Strang splitting can be given as [15]

$$\begin{pmatrix} u^{[n+1]} \\ v^{[n+1]} \end{pmatrix} = \mathcal{S}_\tau \begin{pmatrix} u^{[n]} \\ v^{[n]} \end{pmatrix} = \chi_T^{\tau/2} \circ \chi_V^\tau \circ \chi_T^{\tau/2} \begin{pmatrix} u^{[n]} \\ v^{[n]} \end{pmatrix}, \quad n = 0,1,\ldots, \tag{4.20}$$

with $u^{[0]} = \phi$ and $v^{[0]} = \gamma$. In fact, it is easy to verify that (4.5), (4.6), (4.7) and (4.10) are equivalent to (4.14), (4.15), (4.16) and (4.19), respectively. Thus it is straightforward to get that (4.12) is equivalent to (4.20), and (4.13) is the same as (4.20).

Remark 12. Another second-order semi-discretization of the relativistic NLSE (4.2) can be given as

$$\psi^{[n+1]} = \varphi_V^{\tau/2} \circ \varphi_T^\tau \circ \varphi_V^{\tau/2}(\psi^{[n]}), \qquad n = 0,1,2,\ldots, \tag{4.21}$$

which can immediately generate a semi-discretization of the NKGE (1.5) via (4.13). Again, it is easy to check that this discretization is the same as the discretization of the NKGE (4.4) and (1.5) by a similar second-order Strang-type time-splitting as

$$\begin{pmatrix} u^{[n+1]} \\ v^{[n+1]} \end{pmatrix} = \chi_V^{\tau/2} \circ \chi_T^{\tau} \circ \chi_V^{\tau/2} \begin{pmatrix} u^{[n]} \\ v^{[n]} \end{pmatrix}, \qquad n = 0, 1, 2, \ldots. \tag{4.22}$$

Furthermore, the above second-order time-splitting discretization of the NKGE (1.5) is equivalent to a symmetric second-order exponential integrator of Deuflhard-type for discretizing the NKGE (1.5) directly [15].

Let ψ_j^n be the numerical approximation of $\psi(x_j, t_n)$ for $j \in \mathcal{T}_N^0$ and $n \geq 0$ and denote $\psi^n = (\psi_0^n, \psi_1^n, \ldots, \psi_N^n)^T \in \mathbb{C}^{N+1}$ for $n = 0, 1, \ldots$. Then, a time-splitting Fourier pseudospectral (TSFP) method for discretizing the relativistic NLSE (4.2) via (4.12) with a Fourier pseudospectral discretization in space is given as

$$\begin{aligned}
\psi_j^{(1)} &= \sum_{l \in \mathcal{J}_N} e^{i\frac{\tau\zeta_l}{2}} \, \widetilde{(\psi^n)}_l \, e^{i\mu_l(x_j - a)}, \\
\psi_j^{(2)} &= \psi_j^{(1)} + \varepsilon^2 \tau \, F_j^n, \qquad F_j^n = i \sum_{l \in \mathcal{J}_N} \frac{1}{\zeta_l} \widetilde{(B(\psi^{(1)}))}_l \, e^{i\mu_l(x_j - a)}, \\
\psi_j^{n+1} &= \sum_{l \in \mathcal{J}_N} e^{i\frac{\tau\zeta_l}{2}} \, \widetilde{(\psi^{(2)})}_l \, e^{i\mu_l(x_j - a)}, \quad j \in \mathcal{T}_N^0, \quad n = 0, 1, \ldots,
\end{aligned} \tag{4.23}$$

where $\zeta_l = \sqrt{1 + \mu_l^2}$ for $l \in \mathcal{J}_N$, $\psi^{(k)} = (\psi_0^{(k)}, \psi_1^{(k)}, \ldots, \psi_N^{(k)})^T \in \mathbb{C}^{N+1}$ for $k = 1, 2$, $B(\psi^{(1)}) := (B(\psi_0^{(1)}), B(\psi_2^{(1)}), \ldots, B(\psi_N^{(1)}))^T \in \mathbb{R}^{N+1}$ and

$$\psi_j^0 = \phi(x_j) - i \sum_{l \in \mathcal{J}_N} \frac{\widetilde{\gamma}_l}{\zeta_l} e^{i\mu_l(x_j - a)}, \quad j \in \mathcal{T}_N^0.$$

Let u_j^n and v_j^n be the approximations of $u(x_j, t_n)$ and $v(x_j, t_n)$, respectively, for $j \in \mathcal{T}_N^0$ and $n \geq 0$, and denote $u^n = (u_0^n, u_1^n, \ldots, u_N^n)^T \in \mathbb{R}^{N+1}$ and $v^n = (v_0^n, v_1^n, \ldots, v_N^n)^T \in \mathbb{R}^{N+1}$ for $n = 0, 1, \ldots$. Combining (4.13) and (4.23), we obtain the full-discretization of the NKGE (1.5) by the TSFP method as

$$\begin{aligned}
u_j^{n+1} &= \frac{1}{2}\left(\psi_j^{n+1} + \overline{\psi_j^{n+1}}\right), \\
v_j^{n+1} &= \frac{i}{2} \sum_{l \in \mathcal{J}_N} \zeta_l \left[\widetilde{(\psi^{n+1})}_l - \widetilde{(\overline{\psi^{n+1}})}_l\right] e^{i\mu_l(x_j - a)},
\end{aligned} \qquad j \in \mathcal{T}_N^0, \quad n \geq 0, \tag{4.24}$$

with $u_j^0 = \phi(x_j)$ and $v_j^0 = \gamma(x_j)$ for $j \in \mathcal{T}_N^0$.

Specifically, plugging (4.23) into (4.24) or discretizing (4.20) directly in space by the Fourier pseudospectral method, we get the full-discretization of the NKGE (1.5) by the TSFP method (in explicit formulation in the original variable u) as

$$\begin{aligned}
u_j^{(n,1)} &= \mathcal{L}_u\left(\frac{\tau}{2}, u^n, v^n\right)_j, & v_j^{(n,1)} &= \mathcal{L}_v\left(\frac{\tau}{2}, u^n, v^n\right)_j, \\
u_j^{(n,2)} &= u_j^{(n,1)}, & v_j^{(n,2)} &= v_j^{(n,1)} - \tau\varepsilon^2\big(u_j^{(n,1)}\big)^3, \\
u_j^{n+1} &= \mathcal{L}_u\left(\frac{\tau}{2}, u^{(n,2)}, v^{(n,2)}\right)_j, & v_j^{n+1} &= \mathcal{L}_v\left(\frac{\tau}{2}, u^{(n,2)}, v^{(n,2)}\right)_j,
\end{aligned} \tag{4.25}$$

where

$$\left.\begin{aligned}
\mathcal{L}_u(\tau, u, v)_j &= \sum_{l\in\mathcal{T}_N}\left[\cos(\tau\zeta_l)\widetilde{u}_l + \zeta_l^{-1}\sin(\tau\zeta_l)\widetilde{v}_l\right]e^{i\mu_l(x_j-a)}, \\
\mathcal{L}_v(\tau, u, v)_j &= \sum_{l\in\mathcal{T}_N}\left[-\zeta_l\sin(\tau\zeta_l)\widetilde{u}_l + \cos(\tau\zeta_l)\widetilde{v}_l\right]e^{i\mu_l(x_j-a)},
\end{aligned}\right. \quad j \in \mathcal{T}_N^0.$$

The TSFP method (4.25) (or (4.24) with (4.23)) for the NKGE (1.5) is explicit, time symmetric and easy to be extended to higher dimensions. The memory cost of the TSFP method is $O(N)$ and the computational cost per time step is $O(N\ln N)$. In addition, the total cost for the long-time dynamics up to the time $T_\varepsilon = T/\varepsilon^2$ with fixed $T > 0$ is $O\left(\frac{T_\varepsilon N\ln N}{\tau}\right) = O\left(\frac{TN\ln N}{\tau\varepsilon^2}\right)$.

4.2. *Improved uniform error bounds in the long-time regime*

We make the following assumptions on the exact solution $u := u(x,t)$ of the NKGE (1.5) up to the time at $T_\varepsilon = T/\varepsilon^2$ with $T > 0$ fixed:

$$\text{(C)} \qquad \begin{aligned}
&u \in L^\infty\left([0,T_\varepsilon]; H_{\text{per}}^m\right), && \partial_t u \in L^\infty\left([0,T_\varepsilon]; H_{\text{per}}^{m-1}\right), \\
&\|u\|_{L^\infty([0,T_\varepsilon];H_{\text{per}}^m)} \lesssim 1, && \|\partial_t u\|_{L^\infty([0,T_\varepsilon];H_{\text{per}}^{m-1})} \lesssim 1,
\end{aligned}$$

with $m \geq 2$. First, we can establish the following error bounds of the TSFP method [29].

Theorem 13. *Let u^n be the numerical approximation obtained from the TSFP (4.23)–(4.24) or equivalently (4.25). Under the assumption (C), there exist $h_0 > 0$ and $\tau_0 > 0$ sufficiently small and independent of ε such that, for any $0 < \varepsilon \leq 1$, when $0 < h \leq h_0$ and $0 < \tau \leq \tau_0$, we have the following error estimates for $0 \leq n \leq \frac{T/\varepsilon^2}{\tau}$*

$$\|u(\cdot,t_n) - I_N u^n\|_{H^1} + \|\partial_t u(\cdot,t_n) - I_N v^n\|_{L^2} \lesssim h^{m-1} + \tau^2. \tag{4.26}$$

Furthermore, there exists a constant $M_1 > 0$ *depending on* T, $\|\phi\|_{H^m}$, $\|\gamma\|_{H^{m-1}}$, $\|u\|_{L^\infty([0,T_\varepsilon];H^{m-1}_{\rm per})}$ *and* $\|\partial_t u\|_{L^\infty([0,T_\varepsilon];H^{m-1}_{\rm per})}$ *such that the numerical solution satisfies*

$$\|I_N u^n\|_{H^m} + \|I_N v^n\|_{H^{m-1}} \leq M_1, \quad 0 \leq n \leq \frac{T/\varepsilon^2}{\tau}. \tag{4.27}$$

Based on the above error estimates, we introduce the **Regularity Compensation Oscillation** (RCO) technique to establish the following improved uniform error bounds for the Strang splitting method up to the long-time T/ε^2.

Theorem 14. *Let* u^n *be the numerical approximation obtained from the TSFP (4.23)–(4.24) or equivalently (4.25). Under the assumption (C), there exist* $h_0 > 0$ *and* $0 < \tau_0 < 1$ *sufficiently small and independent of* ε *such that, for any* $0 < \varepsilon \leq 1$, *when* $0 < h \leq h_0$ *and* $0 < \tau < \alpha\frac{\pi(b-a)\tau_0}{2\sqrt{\tau_0^2(b-a)^2+4\pi^2(1+\tau_0^2)}}$ *for a fixed constant* $\alpha \in (0,1)$, *we have the following improved uniform error estimates for* $0 \leq n \leq \frac{T/\varepsilon^2}{\tau}$

$$\|u(\cdot,t_n) - I_N u^n\|_{H^1} + \|\partial_t u(\cdot,t_n) - I_N v^n\|_{L^2} \lesssim h^{m-1} + \varepsilon^2\tau^2 + \tau_0^m. \tag{4.28}$$

In particular, if the exact solution is sufficiently smooth, e.g., $u, \partial_t u \in H^\infty$, *the improved uniform error bounds for sufficiently small* τ *is*

$$\|u(\cdot,t_n) - I_N u^n\|_{H^1} + \|\partial_t u(\cdot,t_n) - I_N v^n\|_{L^2} \lesssim h^{m-1} + \varepsilon^2\tau^2. \tag{4.29}$$

Remark 15. The main idea of the RCO approach is choosing a frequency cut-off parameter τ_0 (which is in the error bound (4.28)) and controlling the high frequency modes ($> 1/\tau_0$) by the smoothness of the exact solution and analyze the low frequency modes ($\leq 1/\tau_0$) by phase cancellation and energy method. $\tau_0 \in (0,1)$ is a cut-off parameter introduced in the analysis and the requirement on τ (essentially $\tau \lesssim \tau_0$) enables the improved estimates on the low Fourier modes $|l| < 1/\tau_0$, where the constant in front of $\varepsilon^2\tau^2$ depends on α. The high Fourier modes $|l| > 1/\tau_0$ are treated by Fourier projection. τ_0 can be arbitrary as long as the assumed relation between τ and τ_0 holds, i.e., τ_0 can be fixed, or depend on τ, e.g., $\tau_0 = \frac{2\sqrt{8\pi^2+(b-a)^2}}{\alpha(b-a)\pi}\tau$.

Remark 16. The second-order Strang splitting method is adopted to discretize the NKGE (1.5) and it is straightforward to design the first-order Lie-Trotter splitting method [33] and fourth-order partitioned Runge-Kutta (PRK4) splitting method [34, 35]. Under appropriate assumptions on the exact solution, the improved uniform error bounds can be extended to the first-order Lie-Trotter splitting and the fourth-order partitioned Runge-Kutta splitting method with improved uniform error bounds at $\varepsilon^2\tau$ and $\varepsilon^2\tau^4$, respectively.

4.3. *Extensions to general NKGE*

Consider the following complex NKGE with a general power nonlinearity

$$\begin{cases} \partial_{tt}u(\mathbf{x},t) - \Delta u(\mathbf{x},t) + u(\mathbf{x},t) + \varepsilon^{2p}|u(\mathbf{x},t)|^{2p}u(\mathbf{x},t) = 0, \\ u(\mathbf{x},0) = \phi(\mathbf{x}), \quad \partial_t u(\mathbf{x},0) = \gamma(\mathbf{x}), \quad \mathbf{x} \in \Omega, \end{cases} \tag{4.30}$$

where $u := u(\mathbf{x},t)$ is a complex-valued scalar field, $p \in \mathbb{N}^+$ is the power index, and the initial data $\phi(\mathbf{x})$ and $\gamma(\mathbf{x})$ are two given complex-valued functions which are independent of the parameter ε. The domain Ω and periodic boundary conditions are as in (1.1). The local/global well-posedness and scattering properties of the Cauchy problem (4.30) have been widely studied in the literature and references therein [6, 12, 36–39]. From the analytical results, the life-span of a smooth solution to the complex NKGE (4.30) is at least $O(\varepsilon^{-2p})$.

For simplicity of notations, we only show the numerical scheme and error estimates for the NKGE in 1D under periodic boundary conditions as

$$\begin{cases} \partial_{tt}u(x,t) - \Delta u(x,t) + u(x,t) + \varepsilon^{2p}|u(x,t)|^{2p}u(x,t) = 0, \\ u(x,0) = \phi(x), \quad \partial_t u(x,0) = \gamma(x), \quad x \in \Omega. \end{cases} \tag{4.31}$$

Similarly, introducing new variables $v(x,t) = \partial_t u(x,t)$ and

$$\eta_\pm(x,t) = u(x,t) \mp i\, \langle \nabla \rangle^{-1} v(x,t), \quad x \in \overline{\Omega}, \quad t \geq 0, \tag{4.32}$$

and denoting $f(\varphi) = |\varphi|^{2p}\varphi$, the complex NKGE (4.31) can be reformulated into the following coupled relativistic NLSEs:

$$\begin{cases} i\partial_t \eta_\pm \pm \langle \nabla \rangle \eta_\pm \pm \varepsilon^{2p} \langle \nabla \rangle^{-1} f\left(\dfrac{1}{2}\eta_+ + \dfrac{1}{2}\eta_-\right) = 0, \\ \eta_\pm(t=0) = \phi \mp i\, \langle \nabla \rangle^{-1}\gamma. \end{cases} \tag{4.33}$$

Let $\eta^n_{\pm,j}$ be the approximations of $\eta_\pm(x_j,t_n)$ for $j \in \mathcal{T}^0_N$ and $n \geq 0$, and denote $\eta^n_\pm = (\eta^n_{\pm,0}, \eta^n_{\pm,1}, \ldots, \eta^n_{\pm,N})^T \in \mathbb{C}^{n+1}$ as the solution at $t_n = n\tau$. The second-order time-splitting Fourier pseudospectral (TSFP) discretization for the relativistic NLSE (4.33) is given by

$$\begin{aligned} \eta^{(1)}_{\pm,j} &= \sum_{l \in \mathcal{T}_N} e^{\pm\, i\frac{\tau\varsigma_l}{2}} \, \widetilde{(\eta^n_\pm)}_l \, e^{i\mu_l(x_j-a)}, \\ \eta^{(2)}_{\pm,j} &= \eta^{(1)}_{\pm,j} \pm \varepsilon^{2p}\tau f^n_j, \qquad\qquad j \in \mathcal{T}^0_N, \quad n \geq 0, \\ \eta^{n+1}_{\pm,j} &= \sum_{l \in \mathcal{T}_N} e^{\pm\, i\frac{\tau\varsigma_l}{2}} \, \widetilde{(\eta^{(2)}_\pm)}_l \, e^{i\mu_l(x_j-a)}, \end{aligned} \tag{4.34}$$

with

$$\eta^0_{\pm,j} = \phi(x_j) \mp i \sum_{l\in\mathcal{J}_N} \frac{1}{\zeta_l}\widetilde{\gamma}_l e^{i\mu_l(x_j-a)},$$
$$f^n_j = i \sum_{l\in\mathcal{J}_N} \frac{1}{\zeta_l}\left(\widetilde{f((\eta^n_+ + \eta^n_-)/2)}\right)_l e^{i\mu_l(x_j-a)}. \tag{4.35}$$

Then u_j^{n+1} and v_j^{n+1} for $j \in \mathcal{T}_N^0, n \geq 0$, which are approximations of $u(x_j, t_{n+1})$ and $v(x_j, t_{n+1})$, respectively, can be recovered by

$$u^{n+1}_j = \frac{1}{2}\left(\eta^{n+1}_{+,j} + \eta^{n+1}_{-,j}\right),$$
$$v^{n+1}_j = \frac{i}{2}\sum_{l\in\mathcal{J}_N} \zeta_l\left(\left(\widetilde{\eta^{n+1}_+}\right)_l - \left(\widetilde{\eta^{n+1}_-}\right)_l\right)e^{i\mu_l(x_j-a)}. \tag{4.36}$$

We assume the existence of the exact solution $u := u(x,t)$ of the NKGE (4.31) up to the time $T_{\varepsilon,p} = T/\varepsilon^{2p}$ ($T > 0$ fixed) and

$$\text{(D)} \qquad \begin{aligned} &u \in L^\infty\left([0, T_{\varepsilon,p}]; H^m_{\text{per}}\right), && \partial_t u \in L^\infty\left([0, T_{\varepsilon,p}]; H^{m-1}_{\text{per}}\right),\\ &\|u\|_{L^\infty\left([0,T_{\varepsilon,p}];H^m_{\text{per}}\right)} \lesssim 1, && \|\partial_t u\|_{L^\infty\left([0,T_{\varepsilon,p}];H^{m-1}_{\text{per}}\right)} \lesssim 1,\end{aligned}$$

with $m \geq 2$, then the following improved uniform error bounds for the TSFP method (4.34)–(4.36) can be established up to the time $T_{\varepsilon,p}$.

Theorem 17. *Let u^n and v^n be the numerical approximations obtained from the TSFP (4.34)–(4.36). Under the assumption (D), there exist constants $h_0 > 0$ and $0 < \tau_0 < 1$ sufficiently small and independent of ε such that, for any $0 < \varepsilon \leq 1$, when $0 < h \leq h_0$ and $0 < \tau \leq \alpha\tau_0$ for some fixed constant $\alpha > 0$, we have the following improved uniform error estimates for $0 \leq n \leq \frac{T/\varepsilon^{2p}}{\tau}$*

$$\|u(\cdot, t_n) - I_N u^n\|_{H^1} + \|\partial_t u(\cdot, t_n) - I_N v^n\|_{L^2} \lesssim h^{m-1} + \varepsilon^{2p}\tau^2 + \tau_0^m. \tag{4.37}$$

In particular, if the exact solution is sufficiently smooth, e.g., $u, \partial_t u \in H^\infty$, the improved uniform error bounds for sufficiently small τ is

$$\|u(\cdot, t_n) - I_N u^n\|_{H^1} + \|\partial_t u(\cdot, t_n) - I_N v^n\|_{L^2} \lesssim h^{m-1} + \varepsilon^{2p}\tau^2. \tag{4.38}$$

Remark 18. The NKGE (4.31) conserves the energy as [23, 29]

$$\begin{aligned} E(t) &:= \int_\Omega \left[|\partial_t u(x,t)|^2 + |\nabla u(x,t)|^2 + |u(x,t)|^2 + \frac{\varepsilon^{2p}}{p+1}|u(x,t)|^{2p+2}\right]dx \\ &\equiv \int_\Omega \left[|\gamma(x)|^2 + |\nabla\phi(x)|^2 + |\phi(x)|^2 + \frac{\varepsilon^{2p}}{p+1}|\phi(x)|^{2p+2}\right]dx \\ &= E(0), \quad t \geq 0. \end{aligned}$$

Introducing the discrete energy at $t = t_n$ with the mesh size h as

$$E_h^n = h \sum_{j=0}^{N-1} \left[|v_j^n|^2 + |(\partial_x u)_j^n|^2 + |u_j^n|^2 + \frac{\varepsilon^{2p}}{p+1} |u_j^n|^{2p+2} \right]$$

where

$$(\partial_x u)_j^n = i \sum_{l \in \mathcal{T}_N} \mu_l \widetilde{(u^n)}_l e^{i\mu_l (x_j - a)}, \quad j = 0, 1, \ldots, N-1,$$

we have the following estimates for the discrete energy

$$|E_h^n - E_h^0| \lesssim h^{m-1} + \varepsilon^{2p}\tau^2 + \tau_0^m, \quad 0 \le n \le \frac{T/\varepsilon^{2p}}{\tau}. \tag{4.39}$$

In addition, if the exact solution is sufficiently smooth, e.g., $u, \partial_t u \in H^\infty$, the estimate for the discrete energy for sufficiently small τ is

$$|E_h^n - E_h^0| \lesssim h^{m-1} + \varepsilon^{2p}\tau^2, \quad 0 \le n \le \frac{T/\varepsilon^{2p}}{\tau}. \tag{4.40}$$

4.4. *Numerical results*

In this subsection, we present numerical examples for the TSFP (4.34)–(4.36) to solve the general NKGE (4.31).

4.4.1. *The long-time dynamics in 1D*

First, we test the long-time errors of the TSFP (4.34)–(4.36) for the NKGE (4.31) with $p = 2$ and real-valued initial data as

$$\phi(x) = \frac{3}{2 + \cos^2(x)}, \quad \gamma(x) = \frac{3}{4 + \cos^2(x)}, \quad x \in \Omega = (0, 2\pi). \tag{4.41}$$

The numerical "exact" solution is computed by the TSFP (4.34)–(4.36) with a very fine mesh size $h_e = \pi/60$ and time step $\tau_e = 10^{-4}$. To quantify the error, we introduce the following error functions

$$e_1(t_n) = \|u(x, t_n) - I_N u^n\|_{H^1}, \quad e_{1,\max}(t_n) = \max_{0 \le q \le n} e_1(t_q). \tag{4.42}$$

In the rest of this part, the spatial mesh size is always chosen sufficiently small such that the spatial errors can be neglected when considering the long-time temporal errors.

Figure 1 displays the long-time errors of the TSFP (4.34)–(4.36) for the NKGE (4.31) with $p = 2$, the fixed time step τ and different ε, which confirms the improved uniform error bounds in H^1-norm at $O(\varepsilon^4\tau^2)$ up to time at $O(1/\varepsilon^4)$. Figures 2 and 3 depict the spatial and temporal errors

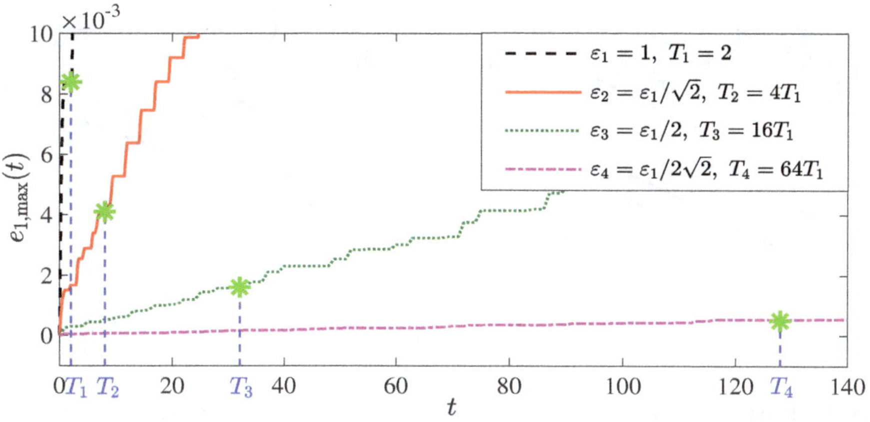

Fig. 1. Long-time temporal errors of the TSFP (4.34)–(4.36) for the NKGE (4.31) with $p = 2$ and different ε.

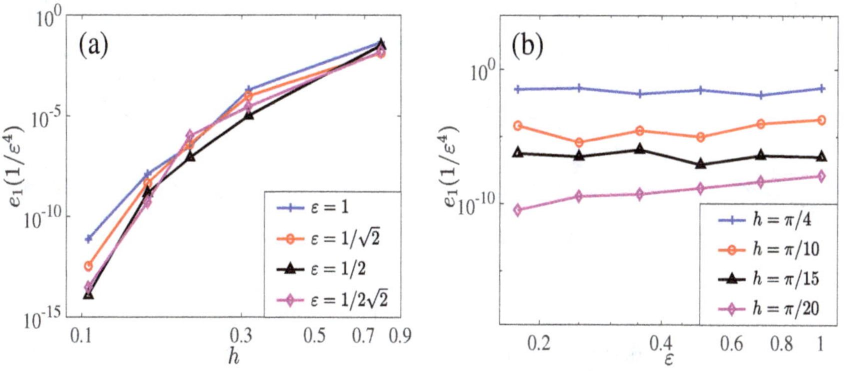

Fig. 2. Long-time spatial errors of the TSFP (4.34)–(4.36) for the NKGE (4.31) with $p = 2$ at $t = 1/\varepsilon^4$.

of the TSFP (4.34)–(4.36) for the NKGE (4.31) with $p = 2$ at $t = 1/\varepsilon^4$, respectively. Figure 2 indicates the spectral accuracy of the TSFP (4.34)–(4.36) for the NKGE (4.31) in space and the spatial errors are independent of the small parameter ε. Each line in Figure 3(a) corresponds to a fixed ε and shows the global errors in H^1-norm versus the time step τ, which confirms the second-order convergence of the TSFP (4.34)–(4.36) for the NKGE (4.31) in time. Figure 3(b) again validates that the global errors in H^1- norm behave like $O(\varepsilon^4\tau^2)$ up to the time at $O(1/\varepsilon^4)$. Figure 4 shows

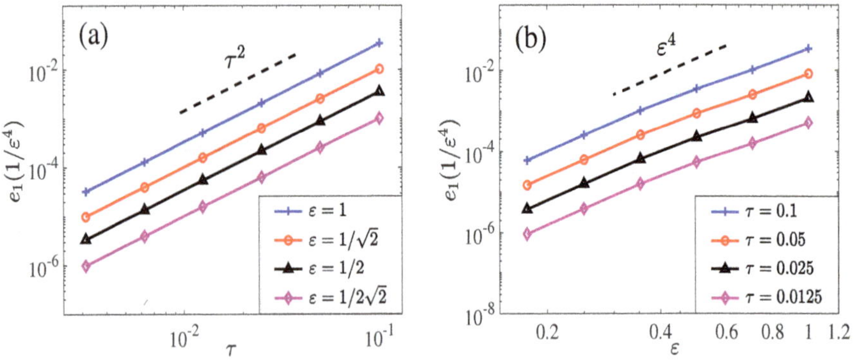

Fig. 3. Long-time temporal errors of the TSFP (4.34)–(4.36) for the NKGE (4.31) with $p = 2$ at $t = 1/\varepsilon^4$.

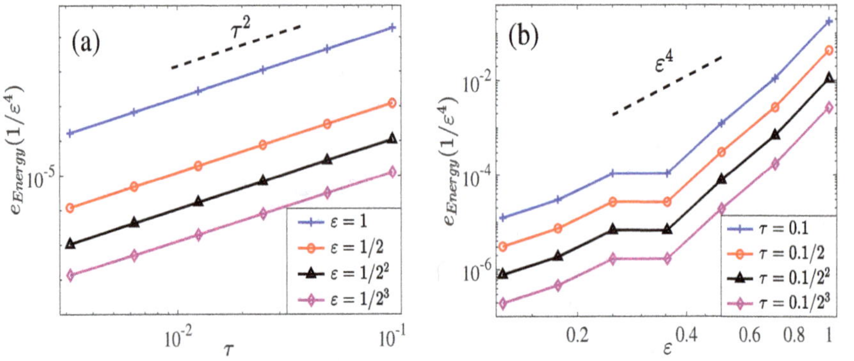

Fig. 4. Long-time errors for the discrete energy of the TSFP (4.34)–(4.36) for the NKGE (4.31) with $p = 2$ at $t = 1/\varepsilon^4$.

the long-time errors for the discrete energy of the TSFP (4.34)–(4.36) for the NKGE (4.31) with $p = 2$ at $t = 1/\varepsilon^4$, which verifies the improved uniform error bounds for the discrete energy (4.39).

For comparisons, we present the temporal errors of the first-, second- and fourth-order splitting methods. Figure 5(a) depicts the temporal errors of three splitting methods with $\varepsilon = 1/2$, which indicates that the higher order splitting method not only has higher order convergence rate but also achieves better accuracy under the same time step size. Figure 5(b) shows the temporal errors of three splitting methods for the fixed time step $\tau = 0.1$ and confirms the improved uniform error bounds for all the three splitting methods up to the time at $O(1/\varepsilon^4)$.

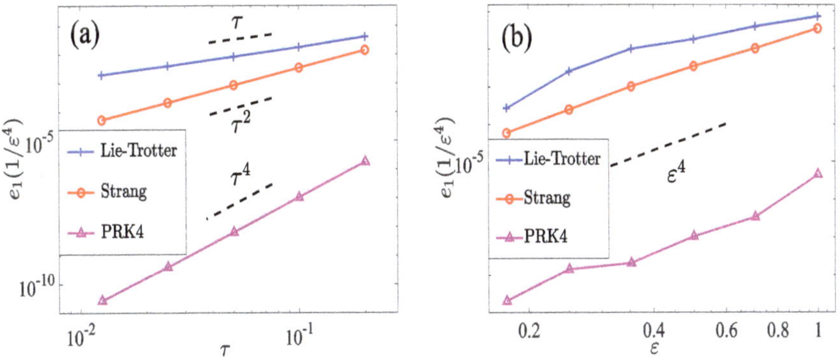

Fig. 5. Comparisons of the first-, second- and fourth-order splitting methods for the NKGE (4.31) with $p = 2$ at $t = 1/\varepsilon^4$: (a) convergence in terms of τ for fixed $\varepsilon = 1/2$; (b) convergence in terms of ε for fixed $\tau = 0.1$.

4.4.2. *The long-time dynamics in 2D*

In this part, we show an example in 2D with the irrational aspect ratio of the domain $(x, y) \in \Omega = (0, 1) \times (0, 2\pi)$. In the numerical experiment, we choose $p = 1$ and the initial data as

$$\phi(x, y) = \frac{2}{1 + \cos^2(2\pi x + y)}, \quad \gamma(x, y) = \frac{3}{2 + 2\cos^2(2\pi x + y)}.$$

Figure 6 presents the long-time errors of the TSFP method for the NKGE in 2D with a fixed time step τ and different ε, which confirms that the improved uniform error bounds at $O(\varepsilon^2\tau^2)$ up to the time at $O(1/\varepsilon^2)$ are also suitable for the domain with irrational aspect ratio. Figure 7 depicts the temporal errors for the TSFP method for the NKGE in 2D at $t = 1/\varepsilon^2$, which again indicates that the TSFP method is second-order in time and validates the improved uniform error bounds up to the time at $O(1/\varepsilon^2)$.

5. Comparisons and extensions

In this section, we present the comparisons of different numerical methods for the NKGE (1.5) in the long-time regime and extensions to an oscillatory

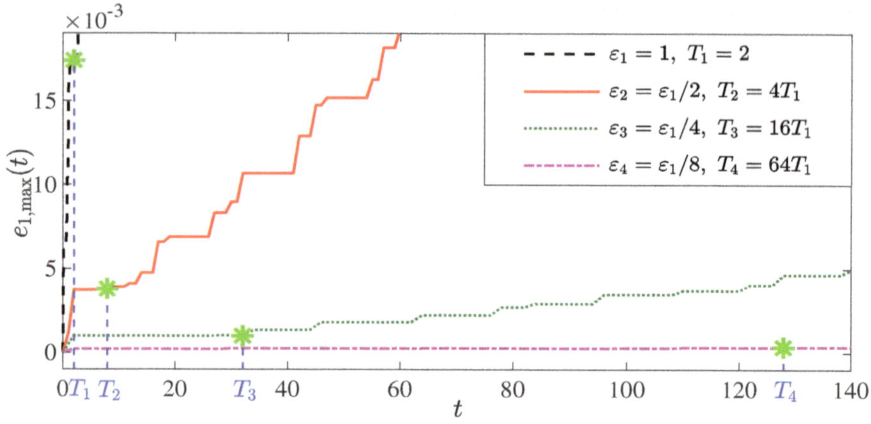

Fig. 6. Long-time temporal errors of the TSFP method for the NKGE in 2D with different ε.

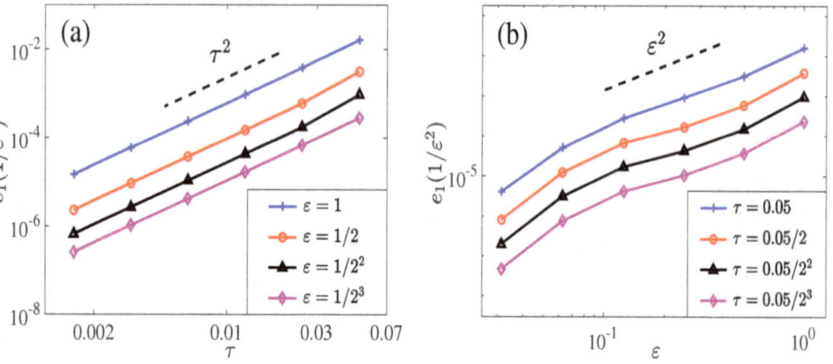

Fig. 7. Long-time temporal errors of the TSFP method for the NKGE in 2D at $t = 1/\varepsilon^2$.

NKGE and other dispersive PDEs including the (nonlinear) Schrödinger equation and (nonlinear) Dirac equation.

5.1. *Comparisons of different numerical methods*

In above sections, we introduce the finite different time domain (FDTD) methods, exponential integrator Fourier pseudospectral (EIFP) method and time-splitting Fourier pseudospectral (TSFP) method to solve the NKGE and rigorously establish the corresponding error bounds in the long-time regime. Figure 8 compares the FDTD method, EIFP method and TSFP method for the long-time dynamics of the NKGE. Figure 8(a) depicts temporal errors of these three methods with $\varepsilon = 1/2$, which indicates that they are all second-order methods for fixed ε. Figure 8(b) shows that just the time-splitting method has improved uniform error bounds at $O(\varepsilon^2)$ for fixed time step τ, while the EIFP method is uniform and the FDTD method performs like $O(1/\varepsilon^2)$ up to the time at $O(1/\varepsilon^2)$. Long-time error bounds of different spatial and temporal discretizations are shown in Table 5. In addition, we get the spatial and temporal resolution for each numerical method in the long-time regime, which can be concluded in Table 6.

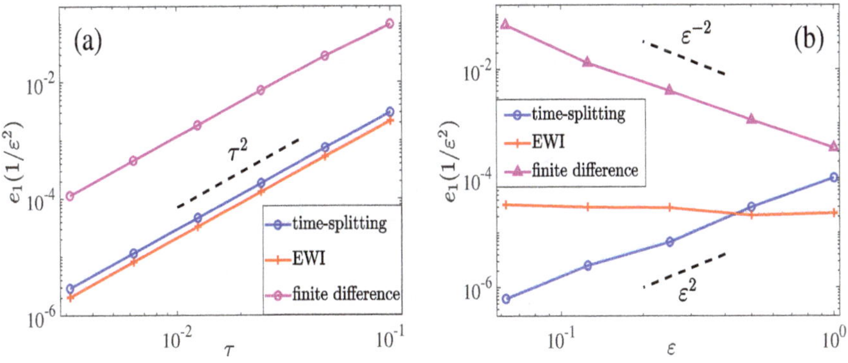

Fig. 8. Comparisons of the FDTD, EIFP and TSFP methods for the NKGE (1.1) in 1D at $t = 1/\varepsilon^2$: (a) convergence in terms of τ for fixed $\varepsilon = 1/2$; (b) convergence in terms of ε for fixed $\tau = 0.01$.

5.2. *Extensions to an oscillatory NKGE*

The numerical methods and error estimates for the long-time dynamics of the NKGE can be extended to an oscillatory NKGE. Introducing a re-scale in time

$$t = \frac{r}{\varepsilon^{2p}} \Leftrightarrow r = \varepsilon^{2p}t, \quad \nu(\mathbf{x}, r) = u(\mathbf{x}, t), \tag{5.1}$$

Table 5. Long-time error bounds of different numerical methods for the NKGE (1.5).

Error bounds: Spatial \ Temporal	FD	EI	TS
2nd FD	$O\left(\frac{h^2}{\varepsilon^2}+\frac{\tau^2}{\varepsilon^2}\right)$	$O\left(\frac{h^2}{\varepsilon^2}+\tau^2\right)$	$O\left(\frac{h^2}{\varepsilon^2}+\varepsilon^2\tau^2\right)$
4cFD	$O\left(\frac{h^4}{\varepsilon^2}+\frac{\tau^2}{\varepsilon^2}\right)$	$O\left(\frac{h^4}{\varepsilon^2}+\tau^2\right)$	$O\left(\frac{h^4}{\varepsilon^2}+\varepsilon^2\tau^2\right)$
Spectral	$O\left(h^m+\frac{\tau^2}{\varepsilon^2}\right)$	$O\left(h^m+\tau^2\right)$	$O\left(h^m+\varepsilon^2\tau^2\right)$

Table 6. Comparison of spatial and temporal resolution of different numerical methods for the NKGE (1.5).

Resolution: Spatial \ Temporal	FD	EI	TS
2nd FD	$h=O(\varepsilon)$ $\tau=O(\varepsilon)$	$h=O(\varepsilon)$ $\tau=O(1)$	$h=O(\varepsilon)$ $\tau=O(1)$
4cFD	$h=O(\varepsilon^{1/2})$ $\tau=O(\varepsilon)$	$h=O(\varepsilon^{1/2})$ $\tau=O(1)$	$h=O(\varepsilon^{1/2})$ $\tau=O(1)$
Spectral	$h=O(1)$ $\tau=O(\varepsilon)$	$h=O(1)$ $\tau=O(1)$	$h=O(1)$ $\tau=O(1)$

the NKGE (4.30) can be reformulated into the following oscillatory complex NKGE

$$\begin{cases} \varepsilon^{2p}\partial_{rr}\nu - \dfrac{1}{\varepsilon^{2p}}\Delta\nu + \dfrac{1}{\varepsilon^{2p}}\nu + |\nu|^{2p}\nu = 0, \quad \mathbf{x}\in\Omega,\ r>0, \\ \nu(\mathbf{x},0)=\phi(\mathbf{x}), \quad \partial_r\nu(\mathbf{x},0)=\dfrac{1}{\varepsilon^{2p}}\gamma(\mathbf{x}), \quad \mathbf{x}\in\Omega. \end{cases} \tag{5.2}$$

The solution of the oscillatory NKGE (5.2) propagates waves with amplitude at $O(1)$, wavelength at $O(1)$ and $O(\varepsilon^{2p})$ in space and time, respectively, and wave velocity at $O(\varepsilon^{-2p})$. Denoting $\mu(\mathbf{x}, r) = \partial_r \nu(\mathbf{x}, r)$ and taking the time step $\kappa = \varepsilon^{2p}\tau$, the error estimates of the FDTD methods, EIFP method and TSFP method for the long-time problem can be extended to the oscillatory complex NKGE (5.2) up to the fixed time T. The error bounds of different numerical methods for the oscillatory NKGE (5.2) are shown in Table 7 with the spatial and temporal resolution presented in Table 8. In space, the Fourier pseudospectral method attain the uniform error bound in terms of ε and optimal spatial resolution for the oscillatory NKGE (5.2). In time, both the exponential integrator and time-splitting methods have the optimal temporal resolution. However, with the same temporal resolution, the time-splitting method has additional ε^{2p} improvement to solve the oscillatory NKGE (5.2).

Table 7. Error bounds of different numerical methods for the oscillatory NKGE (5.2).

Error bounds / Temporal / Spatial	FD	EI	TS
2nd FD	$O\left(\frac{h^2}{\varepsilon^{2p}} + \frac{\kappa^2}{\varepsilon^{6p}}\right)$	$O\left(\frac{h^2}{\varepsilon^{2p}} + \frac{\kappa^2}{\varepsilon^{4p}}\right)$	$O\left(\frac{h^2}{\varepsilon^{2p}} + \frac{\kappa^2}{\varepsilon^{2p}}\right)$
4cFD	$O\left(\frac{h^4}{\varepsilon^{2p}} + \frac{\kappa^2}{\varepsilon^{6p}}\right)$	$O\left(\frac{h^4}{\varepsilon^{2p}} + \frac{\kappa^2}{\varepsilon^{4p}}\right)$	$O\left(\frac{h^4}{\varepsilon^{2p}} + \frac{\kappa^2}{\varepsilon^{2p}}\right)$
Spectral	$O\left(h^m + \frac{\kappa^2}{\varepsilon^{6p}}\right)$	$O\left(h^m + \frac{\kappa^2}{\varepsilon^{4p}}\right)$	$O\left(h^m + \frac{\kappa^2}{\varepsilon^{2p}}\right)$

5.3. *Extensions to other dispersive PDEs*

In this subsection, we extend the numerical methods and error estimates for the nonlinear Klein-Gordon equation in the long-time regime to other dispersive PDEs including the (nonlinear) Schrödinger equation and (nonlinear) Dirac equation.

The (nonlinear) Schrödinger equation arises in various physical phenomena, such as quantum mechanics, Bose-Einstein condensates, laser beam propagation, plasma and particle physics [40–42]. We consider the following

Table 8. Comparison of spatial and temporal resolution of different numerical methods for the oscillatory NKGE (5.2).

Resolution: Spatial \ Temporal	FD	EI	TS
2nd FD	$h = O(\varepsilon^p)$ $\kappa = O(\varepsilon^{3p})$	$h = O(\varepsilon^p)$ $\kappa = O(\varepsilon^{2p})$	$h = O(\varepsilon^p)$ $\kappa = O(\varepsilon^{2p})$
4cFD	$h = O(\varepsilon^{p/2})$ $\kappa = O(\varepsilon^{3p})$	$h = O(\varepsilon^{p/2})$ $\kappa = O(\varepsilon^{2p})$	$h = O(\varepsilon^{p/2})$ $\kappa = O(\varepsilon^{2p})$
Spectral	$h = O(1)$ $\kappa = O(\varepsilon^{3p})$	$h = O(1)$ $\kappa = O(\varepsilon^{2p})$	$h = O(1)$ $\kappa = O(\varepsilon^{2p})$

Schrödinger equation [43]

$$i\partial_t\psi(\mathbf{x},t) = -\Delta\psi(\mathbf{x},t) + \varepsilon V(\mathbf{x})\psi(\mathbf{x},t), \ \mathbf{x} \in \Omega, \ t > 0, \tag{5.3}$$

and the nonlinear Schrödinger equation (NLSE)

$$i\partial_t\psi(\mathbf{x},t) = -\Delta\psi(\mathbf{x},t) \pm \varepsilon^2|\psi(\mathbf{x},t)|^2\psi(\mathbf{x},t), \ \mathbf{x} \in \Omega, \ t > 0, \tag{5.4}$$

with the initial data

$$\psi(\mathbf{x},0) = \psi_0(\mathbf{x}), \quad \mathbf{x} \in \overline{\Omega}, \tag{5.5}$$

where $\Omega = \prod_{i=1}^d (a_i, b_i) \subset \mathbb{R}^d$ $(d = 1, 2, 3)$ is a bounded domain equipped with periodic boundary conditions, $\mathbf{x} = (x_1, \cdots, x_d)^T \in \mathbb{R}^d$ is the spatial coordinate, t is time, $\psi(\mathbf{x},t) \in \mathbb{C}$ is the complex order parameter/wave function, $V(\mathbf{x}) \in \mathbb{R}$ is a given external potential, $\varepsilon \in (0,1]$ is a dimensionless parameter. In the Schrödinger equation (5.3), the amplitude of the potential is characterized by the parameter $\varepsilon \in (0,1]$. In the NLSE (5.4), the strength of the nonlinearity is $O(\varepsilon^2)$ – NLSE with weak nonlinearity – and the dynamics of the NLSE (5.4) with $O(1)$-initial data is equivalent to the NLSE with $O(1)$-nonlinearity and $O(\varepsilon)$-initial data – NLSE with small initial data, e.g., by setting $\phi(\mathbf{x},t) = \varepsilon\psi(\mathbf{x},t)$, the NLSE (5.4) with (5.5) becomes

$$\begin{cases} i\partial_t\phi(\mathbf{x},t) = -\Delta\phi(\mathbf{x},t) \pm |\phi(\mathbf{x},t)|^2\phi(\mathbf{x},t), \ \mathbf{x} \in \Omega, \ t > 0, \\ \phi(\mathbf{x},0) = \varepsilon\psi_0(\mathbf{x}) := \phi_0(\mathbf{x}) = O(\varepsilon), \quad \mathbf{x} \in \overline{\Omega}. \end{cases} \tag{5.6}$$

Different numerical methods can be used to solve the Schrödinger equation (5.3) and rigorous error bounds are carried out up to the time at $O(1/\varepsilon)$ including the FDTD methods, the EIFP method and TSFP method. In particular, by employing the unitary flow property of the TSFP method in L^2-norm, we can prove a uniform error bound at time $t_\varepsilon = t/\varepsilon$ as $C(t)\widetilde{C}(T)(h^m + \tau^2)$ up to $t_\varepsilon \leq T_\varepsilon = T/\varepsilon$ for any $T > 0$ and uniformly for $\varepsilon \in (0, 1]$, while h is the mesh size, τ is the time step, $m \geq 2$ and $\tilde{C}(T)$ (the local error bound) depend on the regularity of the exact solution, and $C(t) = C_0 + C_1 t$ grows at most linearly with respect to t with C_0 and C_1 two positive constants independent of T, ε, h and τ. Correspondingly, the long-time error estimates of these numerical methods can be extended to those for the NLSE (5.4) up to the time at $O(1/\varepsilon^2)$ [43].

The Dirac equation is a relativistic wave equation derived by the British physicist Paul Dirac in 1928 [44]. It plays a fundamental role in relativistic particle physics and is consistent with both the principles of quantum mechanics and the theory of special relativity. The Dirac equation describes the motion of the spin-1/2 massive particles such as electrons and positrons. It fully accounts for the details of the hydrogen spectrum and predicts the existence of a new form of matter, antimatter [45]. We consider the following dimensionless Dirac equation in d dimensions ($d = 1, 2, 3$) [46, 47]

$$i\partial_t \Psi = \left(-i\sum_{j=1}^{d} \alpha_j \partial_j + \beta \right) \Psi + \varepsilon \left(V(t,\mathbf{x}) I_4 - \sum_{j=1}^{d} A_j(t,\mathbf{x}) \alpha_j \right) \Psi, \mathbf{x} \in \Omega, \tag{5.7}$$

where $\Omega = \prod_{i=1}^{d}(a_i, b_i) \subset \mathbb{R}^d$ ($d = 1, 2, 3$) is a bounded domain equipped with periodic boundary conditions, $\varepsilon \in (0, 1]$ is a dimensionless parameter, t is time, $\mathbf{x} = (x_1, \ldots, x_d)^T \in \mathbb{R}^d$ is the spatial coordinate vector, $\partial_j = \frac{\partial}{\partial x_j}$ $(j = 1, 2, 3)$, $\Psi := \Psi(t,\mathbf{x}) = (\psi_1(t,\mathbf{x}), \psi_2(t,\mathbf{x}), \psi_3(t,\mathbf{x}), \psi_4(t,\mathbf{x}))^T \in \mathbb{C}^4$ is the complex-valued spinor wave function. I_n is the $n \times n$ identity matrix for $n \in \mathbb{N}$, $V := V(t,\mathbf{x}) \in \mathbb{R}$ is the electric potential, and $\mathbf{A} := \mathbf{A}(t,\mathbf{x}) = (A_1(t,\mathbf{x}), A_2(t,\mathbf{x}), A_3(t,\mathbf{x}))^T \in \mathbb{R}^3$ stands for the magnetic potential. In addition, the 4×4 matrices $\alpha_1, \alpha_2, \alpha_3$ and β are defined as

$$\alpha_1 = \begin{pmatrix} \mathbf{0} & \sigma_1 \\ \sigma_1 & \mathbf{0} \end{pmatrix}, \quad \alpha_2 = \begin{pmatrix} \mathbf{0} & \sigma_2 \\ \sigma_2 & \mathbf{0} \end{pmatrix}, \quad \alpha_3 = \begin{pmatrix} \mathbf{0} & \sigma_3 \\ \sigma_3 & \mathbf{0} \end{pmatrix}, \quad \beta = \begin{pmatrix} I_2 & \mathbf{0} \\ \mathbf{0} & -I_2 \end{pmatrix},$$

where σ_j $(j = 1, 2, 3)$ are the Pauli matrices

$$\sigma_1 = \begin{pmatrix} 0 & 1 \\ 1 & 0 \end{pmatrix}, \quad \sigma_2 = \begin{pmatrix} 0 & -i \\ i & 0 \end{pmatrix}, \quad \sigma_3 = \begin{pmatrix} 1 & 0 \\ 0 & -1 \end{pmatrix}. \tag{5.8}$$

In addition, the dimensionless nonlinear Dirac equation (NLDE) with weak nonlinearity for the wave function $\Psi := \Psi(t, \mathbf{x}) = (\psi_1(t, \mathbf{x}), \psi_2(t, \mathbf{x}), \psi_3(t, \mathbf{x}), \psi_4(t, \mathbf{x}))^T \in \mathbb{C}^4$ is

$$i\partial_t \Psi = \left(-i \sum_{j=1}^{3} \alpha_j \partial_j + \beta \right) \Psi + \varepsilon^2 \mathbf{F}(\Psi)\Psi, \quad \mathbf{x} \in \Omega, \tag{5.9}$$

where

$$\mathbf{F}(\Psi) = \lambda_1(\Psi^* \beta \Psi)\beta + \lambda_2 |\Psi|^2 I_4, \quad \Psi \subset \mathbb{C}^4, \tag{5.10}$$

with $\lambda_1, \lambda_2 = O(1)$ two dimensionless constants for the interaction strength. The weakly interacting regime has been adopted to study the Bose-Einstein condensate of ^{87}Rb.

In order to study the dynamics of the Dirac equation (5.7) and the NLDE (5.9), the initial condition is taken as

$$\Psi(t = 0, \mathbf{x}) = \Psi_0(\mathbf{x}), \quad \mathbf{x} \in \overline{\Omega}.$$

Different spatial and temporal discretizations are applied to solve the Dirac equation (5.7) with rigorous long-time error bounds established up to the time at $O(1/\varepsilon)$ [47, 48]. In space, the finite difference method and spectral method are adopted to discretize the Dirac equation (5.7). In time, the finite difference method, exponential integrator and time-splitting method are considered. For the nonlinear case, long-time error estimates of the proposed numerical schemes can be extended to those for the NLDE (5.9) up to the time at $O(1/\varepsilon^2)$ [49].

6. Conclusion

In this chapter, we introduce various numerical methods for the long-time dynamics of the nonlinear Klein-Gordon equation (NKGE) with weak nonlinearity or small initial data. In space, finite difference method and Fourier spectral method are used to discretize the NKGE. The finite difference method, exponential integrator and time-splitting method are applied to discretize the NKGE in time. For each numerical method, the error bounds with spatial and temporal resolutions are established in the long-time regime. Plentiful numerical examples are presented to confirm the error bounds and comparisons of the numerical methods are concluded, which are really useful for the long-time simulations of the NKGE. In addition, the long-time error estimates of the proposed numerical schemes can be extended to other dispersive PDEs.

Acknowledgments

This work was partially supported by the Ministry of Education of Singapore under its AcRF Tier 2 funding grant MOE-T2EP20122-0002 (A-8000962-00-00) (W. Bao) and the National Natural Science Foundation of China No. 12171041 (Y. Cai), No. 11771036 (Y. Cai) and No. 12401539 (Y. Feng).

References

[1] J. J. Sakurai, *Advanced Quantum Mechanics*. Addison-Wesley (1967).

[2] W. Strauss and L. Vazquez, Numerical solution of a nonlinear Klein-Gordon equation, *Journal of Computational Physics*. **28**(2), 271–278 (1978).

[3] M. Faccioli and L. Salasnich, Spontaneous symmetry breaking and higgs mode: comparing Gross-Pitaevskii and nonlinear Klein-Gordon equations, *Symmetry*. **10**(4), 80 (2018).

[4] V. V. Konotop, A. Sànchez, and L. Vàzquez, Kink dynamics in the weakly stochastic φ^4 model, *Physical Review B*. **44**(6), 2554 (1991).

[5] S. Chikwendu and C. Easwaran, Multiple-scale solution of initial-boundary value problems for weakly nonlinear wave equations on the semi-infinite line, *SIAM Journal on Applied Mathematics*. **52**(4), 946–958 (1992).

[6] S. Klainerman, Global existence of small amplitude solutions to nonlinear Klein–Gordon equations in four space-time dimensions, *Communications on Pure and Applied Mathematics*. **38**, 631–641 (1985).

[7] H. Lindblad, On the lifespan of solutions of nonlinear wave equations with small initial data, *Communications on Pure and Applied Mathematics*. **43** (4), 445–472 (1990).

[8] H. Sunagawa, On global small amplitude solutions to systems of cubic nonlinear Klein–Gordon equations with different mass terms in one space dimension, *Journal of Differential Equations*. **192**(2), 308–325 (2003).

[9] W. Bao and X. Dong, Analysis and comparison of numerical methods for the Klein–Gordon equation in the nonrelativistic limit regime, *Numerische Mathematik*. **120**(2), 189–229 (2012).

[10] D. Cohen, E. Hairer, and C. Lubich, Conservation of energy, momentum and actions in numerical discretizations of non-linear wave equations, *Numerische Mathematik*. **110**, 113–143 (2008).

[11] P. Brenner and W. von Wahl, Global classical solutions of nonlinear wave equations, *Mathematische Zeitschrift*. **176**(1), 87–121 (1981).

[12] K. Ono, Global existence and asymptotic behavior of small solutions for semilinear dissipative wave equations, *Discrete and Continuous Dynamical Systems*. **9**(3), 651–662 (2003).

[13] W. von Wahl, Regular solutions of initial-boundary value problems for linear and nonlinear wave-equations. II, *Mathematische Zeitschrift*. **142**(2), 121–130 (1975).

[14] W. Bao and X. Zhao, Comparison of numerical methods for the nonlinear Klein-Gordon equation in the nonrelativistic limit regime, *Journal of Computational Physics.* **398**, 108886 (2019).

[15] X. Dong, Z. Xu, and X. Zhao, On time-splitting pseudospectral discretization for nonlinear Klein-Gordon equation in nonrelativistic limit regime, *Communications in Computational Physics.* **16**(2), 440–466 (2014).

[16] P. Chartier, N. Crouseilles, M. Lemou, and F. Méhats, Uniformly accurate numerical schemes for highly oscillatory Klein–Gordon and nonlinear Schrödinger equations, *Numerische Mathematik.* **129**, 211–250 (2015).

[17] J. Bernier, E. Faou, and B. Grébert, Long time behavior of the solutions of NLW on the d-dimensional torus, *Forum of Mathematics, Sigma.* **8**, e12 (2020).

[18] J.-M. Delort, On long time existence for small solutions of semi-linear Klein-Gordon equations on the torus, *Journal d'Analyse Mathématique.* **107**(1), 161–194 (2009).

[19] J.-M. Delort and J. Szeftel, Long-time existence for small data nonlinear Klein-Gordon equations on tori and spheres, *International Mathematics Research Notices.* **37**, 1897–1966 (2004).

[20] D. Fang and Q. Zhang, Long-time existence for semi-linear Klein–Gordon equations on tori, *Journal of Differential Equations.* **249**(1), 151–179 (2010).

[21] D. Cohen, E. Hairer, and C. Lubich, Long-time analysis of nonlinearly perturbed wave equations via modulated Fourier expansions, *Archive for Rational Mechanics and Analysis.* **187**, 341–368 (2008).

[22] E. Hairer and C. Lubich, Spectral semi-discretisations of weakly non-linear wave equations over long times, *Foundations of Computational Mathematics.* **8**, 319–334 (2008).

[23] W. Bao, Y. Feng, and W. Yi, Long time error analysis of finite difference time domain methods for the nonlinear Klein-Gordon equation with weak nonlinearity, *Communications in Computational Physics.* **26**(5), 1307–1334 (2019).

[24] Y. Feng, Long time error analysis of the fourth-order compact finite difference methods for the nonlinear Klein–Gordon equation with weak nonlinearity, *Numerical Methods for Partial Differential Equations.* **37**(1), 897–914 (2021).

[25] Y. Feng and W. Yi, Uniform error bounds of an exponential wave integrator for the long-time dynamics of the nonlinear Klein–Gordon equation, *Multiscale Modeling & Simulation.* **19**(3), 1212–1235 (2021).

[26] J. Shen, T. Tang, and L.-L. Wang, *Spectral Methods: Algorithms, Analysis and Applications.* Springer Science & Business Media (2011).

[27] W. Gautschi, Numerical integration of ordinary differential equations based on trigonometric polynomials, *Numerische Mathematik.* **3**, 381–397 (1961).

[28] G. D. Smith, *Numerical Solution of Partial Differential Equations: Finite Difference Methods.* Clarendon Press (1986).

[29] W. Bao, Y. Feng, and C. Su, Uniform error bounds of time-splitting spectral methods for the long-time dynamics of the nonlinear Klein–Gordon

equation with weak nonlinearity, *Mathematics of Computation.* **91**(334), 811–842 (2022).
[30] W. Bao, Y. Cai, and Y. Feng, Improved uniform error bounds on time-splitting methods for long-time dynamics of the nonlinear Klein–Gordon equation with weak nonlinearity, *SIAM Journal on Numerical Analysis.* **60** (4), 1962–1984 (2022).
[31] C. Lubich, On splitting methods for Schrödinger-Poisson and cubic nonlinear Schrödinger equations, *Mathematics of Computation.* **77**(264), 2141–2153 (2008).
[32] C. Su and X. Zhao, On time-splitting methods for nonlinear Schrödinger equation with highly oscillatory potential, *ESAIM: Mathematical Modelling and Numerical Analysis.* **54**(5), 1491–1508 (2020).
[33] H. F. Trotter, On the product of semi-groups of operators, *Proceedings of the American Mathematical Society.* **10**(4), 545–551 (1959).
[34] S. Blanes and P. C. Moan, Practical symplectic partitioned Runge–Kutta and Runge–Kutta–Nyström methods, *Journal of Computational and Applied Mathematics.* **142**(2), 313–330 (2002).
[35] S. Geng, Symplectic partitioned Runge-Kutta methods, *Journal of Computational Mathematics.* **11**, 365–372 (1993).
[36] J. Ginibre and G. Velo, The global Cauchy problem for the non linear Klein-Gordon equation, *Mathematische Zeitschrift.* **189**, 487–505 (1985).
[37] C. S. Morawetz and W. A. Strauss, Decay and scattering of solutions of a nonlinear relativistic wave equation, *Communications on Pure and Applied Mathematics.* **25**(1), 1–31 (1972).
[38] H. Pecher, Nonlinear small data scattering for the wave and Klein-Gordon equation, *Mathematische Zeitschrift.* **185**, 261–270 (1984).
[39] T. Tao, *Local and global analysis of nonlinear dispersive and wave equations.* American Mathematical Society Providence, RI (2006).
[40] X. Antoine, W. Bao, and C. Besse, Computational methods for the dynamics of the nonlinear Schrödinger/Gross–pitaevskii equations, *Computer Physics Communications.* **184**(12), 2621–2633 (2013).
[41] T. Cazenave, *Semilinear Schrödinger Equations.* American Mathematical Society (2003).
[42] C. Sulem and P.-L. Sulem, *The Nonlinear Schrödinger Equation: Self-Focusing and Wave Collapse.* Springer (1999).
[43] W. Bao, Y. Cai, and Y. Feng, Improved uniform error bounds on the time-splitting methods for the long-time (nonlinear) Schrödinger equation, *Mathematics of Computation.* **92**(341), 1109–1139 (2023).
[44] P. A. M. Dirac, *The Principles of Quantum Mechanics.* Oxford University Press (1988).
[45] C. D. Anderson, The positive electron, *Physical Review.* **43**, 491–498 (1933).
[46] W. Bao, Y. Feng, and J. Yin, Improved uniform error bounds on time-splitting methods for the long-time dynamics of the Dirac equation with small potential, *Multiscale Modeling & Simulation.* **20**(3), 1040–1062 (2022).

[47] Y. Feng and J. Yin, Spatial resolution of different discretizations over long-time for the Dirac equation with small potentials, *Journal of Computational and Applied Mathematics.* **412**, 114342 (2022).

[48] Y. Feng, Z. Xu, and J. Yin, Uniform error bounds of exponential wave integrator methods for the long-time dynamics of the Dirac equation with small potentials, *Applied Numerical Mathematics.* **172**, 50–66 (2022).

[49] W. Bao, Y. Cai, and Y. Feng, Improved uniform error bounds on time-splitting methods for the long-time dynamics of the weakly nonlinear Dirac equation, *IMA Journal of Numerical Analysis.* **44**(2), 654–679 (2024).

https://doi.org/10.1142/9789819816620_0003

Chapter 3

Splitting Methods with Complex Coefficients for Linear and Nonlinear Evolution Equations

Sergio Blanes*, Fernando Casas†, Cesáreo González‡ and Mechthild Thalhammer§

*Universitat Politècnica de València, Instituto de Matemática Multidisciplinar, 46022 Valencia, Spain
serblaza@imm.upv.es

†Universitat Jaume I, IMAC and Departament de Matemàtiques, 12071 Castellón, Spain
fernando.casas@uji.es

‡Universidad de Valladolid, Departamento de Matemática Aplicada, 47011 Valladolid, Spain
ome@am.uva.es

§Universität Innsbruck, Institut für Mathematik, 6020 Innsbruck, Austria
mechthild.thalhammer@uibk.ac.at

This contribution is dedicated to the exploration of exponential operator splitting methods for the time integration of evolution equations. It entails the review of previous achievements as well as the depiction of novel results. The standard class of splitting methods involving real coefficients is contrasted with an alternative approach that relies on the incorporation of complex coefficients. In view of long-term computations for linear evolution equations, it is expedient to distinguish symmetric, symmetric-conjugate, and alternating-conjugate schemes. The scope of applications comprises high-order reaction-diffusion equations and complex Ginzburg–Landau equations, which are of relevance in the theories of patterns and superconductivity. Time-dependent Gross–Pitaevskii equations and their parabolic counterparts, which model the dynamics of Bose Einstein condensates and arise in ground state computations, are formally included as special cases. Numerical experiments confirm the validity of theoretical stability conditions and global error bounds as well as the benefits of higher-order complex splitting methods in comparison with standard schemes.

Contents

1. Introduction

Splitting methods. Exponential operator splitting methods constitute nowadays a popular class of numerical algorithms for the integration in time of initial value problems

$$\begin{cases} u'(t) = F\big(u(t)\big)\,, & t \in [t_0, T]\,, \\ u(t_0) \text{ given}\,, \end{cases} \tag{1.1a}$$

involving ordinary or partial differential equations, respectively. A fundamental premise is that the defining function naturally decomposes into two or more parts

$$F = \sum_{\ell=1}^{L} F_\ell\,, \tag{1.1b}$$

so that the problem under consideration can be subdivided into different initial value problems that are easier to solve than the original problem. For detailed information on splitting methods, we refer to [1–3] and the recent review [4].

Benefits. In many relevant instances, exponential operator splitting methods are straightforward to implement and present favourable properties concerning error propagation. In addition, they preserve a variety of structural properties possessed by the differential equation. As a consequence, splitting methods are widely used in computational quantum mechanics when dealing with problems requiring to solve time-dependent Schrödinger equations, see [5–8] as well as [9–19] and references given therein.

Limitations. However, the applicability of exponential operator splitting methods to non-reversible systems has been more limited, see for example [20, 21]. One important reason for that is related with an essential feature of this class of time integration methods: splitting methods necessarily involve negative coefficients when their order is three or higher, see [22, 23]. As a consequence, severe instabilities arise when higher-order splitting methods are applied to evolution equations of parabolic type such as reaction-diffusion equations.

Alternative approach. An alternative to deal with higher-order exponential operator splitting methods for equations evolved by semigroups was proposed in [24–26], see also [27–30]: using instead *complex* coefficients having *positive real part*. In that way, the stability is recovered in certain situations, amongst others, for reaction-diffusion systems involving real constants. Moreover, increasing the order of the scheme typically improves its efficiency over a wide range of accuracies, in contrast with what takes place with real coefficients, see [31]. In view of these appealing properties, the practical implementation and systematic exploration of exponential operator splitting methods with complex coefficients has been carried out, not only by designing new high-order schemes but also by analysing their preservation properties, for both ordinary and partial differential equations [32–36].

Additional features. When considering standard exponential operator splitting methods with real coefficients, left-right palindromic or so-called symmetric compositions are usually preferred, since their construction is easier and in addition they provide time-symmetric approximations to the exact solution. It turns out, however, that in the complex case other alternatives are more favourable with respect to the preservation of properties. The benefits of so-called symmetric-conjugate schemes which are symmetric in the real part and antisymmetric in the imaginary part have been illustrated in [33, 34] for linear unitary problems and in [36] for parabolic equations. Furthermore, it is promising to study complex splitting methods that are obtained by concatenating a given composition with its complex conjugate. Work in progress [37] indicates that schemes of this kind exhibit an excellent behaviour in long-term computations of linear evolution equations.

Objective. In this contribution, our main objective is to analyse the performance of high-order exponential operator splitting methods involving complex coefficients for a general model problem that is defined by a linear combination of powers of the Laplace operator, a space-dependent function, and a nonlinear multiplication operator. We deduce stability conditions and state a convergence result which generalises our former analysis of standard real splitting methods within the context of linear and nonlinear Schrödinger equations, see [17, 18]. The scope of applications includes reaction-diffusion and Ginzburg–Landau-type equations, which are of relevance, amongst others, in the theories of patterns and superconductivity. More precisely, specifically designed high-order reaction-diffusion equations reveal the formation of quasicrystalline patterns, and complex Ginzburg–Landau equations describe the dynamical behaviour of generic spatially extended systems undergoing a supercritical Hopf bifurcation from a stationary to an oscillatory state, see [38, 39]. Time-dependent Gross–Pitaevskii equations modelling the dynamics of Bose–Einstein condensates and their parabolic counterparts arising in ground and excited state computations are formally retained as particular cases, see [5, 7–9].

Outline. This contribution has the following structure. In Section 2, we introduce a general framework of abstract evolution equations and specify the format of complex exponential operator splitting methods. In Section 3, we detail our model problem including high-order reaction-diffusion equations, complex Ginzburg–Landau equations, time-dependent Gross–Pitaevskii equations, and their parabolic counterparts as particular instances. Important means for a stability and error analysis of high-order exponential operator splitting methods are summarised in Section 4. Hereby, it is practical to include instructions along the lines of [17], where standard real splitting methods have been studied for linear Schrödinger equations comprising the Laplace operator and a regular space-dependent function. In essence, the extension to the significantly more involved nonlinear case relies on the powerful calculus of Lie derivatives, see [40]. The rigorous treatment of full discretisations combining time-splitting methods and spectral space discretisations requires suitable adaptations of [18]. Numerical experiments that confirm the theoretical statements and the advantages of high-order complex splitting methods in comparison with standard schemes are finally presented in Section 5.

2. Splitting methods

In this section, we collect basic information on exponential operator splitting methods. For a thorough description of the analytical frameworks and a thematic classification of splitting methods within the field of geometric integration, we refer to [27–30] and [1–4].

Abstract evolution equations. The starting point of our considerations are initial value problems for abstract evolution equations that can be cast into the general format

$$\begin{cases} u'(t) = F\big(u(t)\big) = F_1\big(u(t)\big) + F_2\big(u(t)\big)\,, & t \in [t_0, T]\,, \\ u(t_0) \text{ given}\,, \end{cases} \tag{2.1}$$

see also (1.1). Throughout, we denote by $(X, \|\cdot\|_X)$ the underlying Banach spaces with associated norms and assume that the domains of the defining operators $F_\ell : D(F_\ell) \subseteq X \to X$, $\ell \in \{1, 2\}$, have non-empty intersections. Within the context of nonlinear evolution equations of parabolic or Schrödinger type, it is expedient to employ the frameworks of analytical semigroups or strongly continuous unitary groups, respectively.

Splitting approach. Exponential operator splitting methods for an evolution equation of the form (2.1) rely on the presumption that the numerical approximation of the associated subproblems

$$\begin{cases} u_j'(t) = F_j\big(u_j(t)\big)\,, & t \in [t_0, T]\,, \\ u_j(t_0) \text{ given}\,, \end{cases} \qquad j \in \{1, 2\}\,, \tag{2.2}$$

is significantly simpler compared to the numerical approximation of the original problem.

Scope of applications. The scope of applications includes Hamiltonian systems from classical mechanics and Schrödinger equations from quantum mechanics, where the advantages of exponential operator splitting methods constituting geometric numerical integrators become apparent. It naturally extends to non-reversible systems such as reaction-diffusion systems and Ginzburg–Landau-type equations, which form beautiful spatio-temporal patterns and attract interest, amongst others, in biology, chemistry, geology, and physics. Moreover, the fundamental concepts are applicable to higher-order damped wave equations from nonlinear acoustics and kinetic equations from plasma physics. See for instance [4, 38, 39, 41–43] and references given therein.

Model problems. We study a concrete model problem that comprises a linear combination of powers of the Laplace operator, a space-dependent function, and a nonlinear multiplication operator. Even though the qualitative characteristics of solutions are different, high-order reaction-diffusion equations describing quasicrystals as well as complex Ginzburg–Landau equations arising in superconductivity are formally included. Time-dependent Gross–Pitaevskii equations modelling the dynamics of Bose–Einstein-condensates and their parabolic counterparts emerging in ground and excited state computations by imaginary time propagation are retained as special cases. The reductions to related linear cases lead to partial differential equations that involve, on the one hand, linear differential operators and, on the other hand, multiplication operators defined by space-dependent functions.

Evolution operators. Henceforth, we use a symbolic notation for the exact evolution operators associated with the original problem (2.1) or the subproblems (2.2), respectively, and indicate the dependencies on the current times, the defining operators, and the initial states. For instance, we set

$$\mathcal{E}_{t,F}\big(u(t_0)\big) = u(t)\,, \quad t \in [t_0, T]\,,$$

for the evolution operator corresponding to (2.1).

Iterated commutators. Under the requirement of suitably restricted domains, the commutator of two nonlinear operators is defined by

$$\big[F_1, F_2\big](v) = F_1'(v)\,F_2(v) - F_2'(v)\,F_1(v)\,. \tag{2.3a}$$

More generally, the iterated commutators are determined recursively

$$\mathrm{ad}_{F_1}^{\ell}(F_2) = \begin{cases} F_2\,, & \ell = 0\,, \\ \big[F_1, \mathrm{ad}_{F_1}^{\ell-1}(F_2)\big]\,, & \ell \in \mathbb{N}_{\geq 1}\,. \end{cases} \tag{2.3b}$$

Complex splitting methods. We employ a usual time-stepping approach, denoting by

$$t_0 < t_1 < \cdots < t_N = T\,, \quad \tau_n = t_{n+1} - t_n\,, \quad n \in \{0, 1, \dots, N-1\}\,, \tag{2.4a}$$

the underlying time grid points with corresponding time stepsizes. The recurrence relations for the exact and numerical solution values read as

$$u_{n+1} = \mathcal{S}_{\tau_n,F}(u_n) \approx u(t_{n+1}) = \mathcal{E}_{\tau_n,F}\big(u(t_n)\big),\ n \in \{0, 1, \dots, N-1\}. \tag{2.4b}$$

For suitable choices of the defining method coefficients, complex exponential operator splitting methods can be written as

$$\begin{gathered}\mathcal{S}_{\tau_n,F} = \mathcal{E}_{\tau_n,b_s F_2} \circ \mathcal{E}_{\tau_n,a_s F_1} \circ \cdots \circ \mathcal{E}_{\tau_n,b_1 F_2} \circ \mathcal{E}_{\tau_n,a_1 F_1}\,, \\ (a_j, b_j)_{j=1}^{s} \in \mathbb{C}^{2s}\,.\end{gathered} \tag{2.4c}$$

Evidently, standard splitting methods are included as special cases with real coefficients.

Order and error estimates. Within the context of nonstiff differential equations with sufficiently regular solutions, the order $p \in \mathbb{N}_{\geq 1}$ of a complex exponential operator splitting method (2.4) is defined through the relation

$$\mathcal{L}_{\tau_n,F}(v) = \mathcal{S}_{\tau_n,F}(v) - \mathcal{E}_{\tau_n,F}(v) = \mathcal{O}\big(\tau_n^{p+1}\big) \quad \text{as} \quad \tau_n \searrow 0\,. \tag{2.5}$$

Global error estimates. In view of a rigorous convergence analysis of exponential operator splitting methods, we employ a standard argument based on the telescopic identity to conclude that the validity of stability bounds combined with suitable local error expansions ensuring (2.5) implies global error estimates

$$\begin{gathered}\big\|u_n - u(t_n)\big\|_X \leq C\left(\big\|u_0 - u(t_0)\big\|_X + \tau_{\max}^p\right), \\ n \in \{1, \dots, N\}\,, \quad \tau_{\max} = \max_{n \in \{0,1,\dots,N-1\}} \tau_n\,.\end{gathered}$$

In order to subsume our modus operandi, we meanwhile restrict ourselves to evolution equations (2.1) that are defined by bounded linear operators

$$F_1, F_2 : X \longrightarrow X$$

and complex splitting methods (2.4) applied with constant time stepsizes

$$\tau_n = \tau\,, \quad n \in \{0, 1, \dots, N-1\}\,,$$

since then the exact and numerical solutions can be represented by exponential series

$$\mathcal{E}_{\tau,F} = \mathrm{e}^{\tau\,(F_1+F_2)}\,, \quad \mathcal{S}_{\tau,F} = \mathrm{e}^{b_s \tau F_2}\,\mathrm{e}^{a_s \tau F_1} \cdots \mathrm{e}^{b_1 \tau F_2}\,\mathrm{e}^{a_1 \tau F_1}\,.$$

Consequently, stability bounds follow at once from

$$\begin{gathered}\big\|\mathcal{E}_{\tau,F}\big\|_{X \leftarrow X} \leq \mathrm{e}^{C\tau}\,, \quad \big\|\mathcal{S}_{\tau,F}\big\|_{X \leftarrow X} \leq \mathrm{e}^{C\tau}\,, \\ C = \max\left\{\|F_1\|_{X \leftarrow X} + \|F_2\|_{X \leftarrow X}, \sum_{j=1}^{s}\Big(|a_j|\|F_1\|_{X \leftarrow X} + |b_j|\|F_2\|_{X \leftarrow X}\Big)\right\}.\end{gathered}$$

Provided that the method coefficients satisfy certain order conditions such that local error expansions (2.5) hold true, the desired global error estimates are obtained by means of the telescopic identity

$$u_n - u(t_n) = \big(\mathcal{S}_{\tau,F}\big)^n \,(u_0 - u(t_0)) + \sum_{\ell=0}^{n-1} \big(\mathcal{S}_{\tau,F}\big)^{n-1-\ell} \mathcal{L}_{\tau,F} \,\big(\mathcal{E}_{\tau,F}\big)^{\ell}\, u(t_0)\,,$$
$$n \in \{1, \dots, N\}\,,$$

and straightforward estimation

$$\big\|u_n - u(t_n)\big\|_X \leq \mathrm{e}^{C t_n} \Big(\big\|u_0 - u(t_0)\big\|_X + n\,\big\|\mathcal{L}_{\tau,F}\big\|_{X \leftarrow X}\big\|u(t_0)\big\|_X\Big)\,,$$
$$n \in \{1, \dots, N\}\,.$$

Theorem 1 given in Section 4 below constitutes the natural generalisation to abstract evolution equations involving unbounded nonlinear operators.

Additional features. Whenever the primary interest of the investigations concerns the specification of exponential operator splitting methods, it is convenient to suppress the dependencies on the time stepsizes as well as the defining operators and to replace (2.4c) by the compact representation

$$\mathcal{S} = \big(b_s, a_s, \dots, b_1, a_1\big)\,. \tag{2.6}$$

As special patterns in the distribution of the method coefficients are of importance within the context of long-term computations, we distinguish symmetric (s), symmetric-conjugate (sc), and alternating-conjugate (ac) schemes. Assuming for a moment that the defining operators correspond to real symmetric matrices $F_1, F_2 \in \mathbb{R}^{M \times M}$, these properties can be connected to complex conjugation and transposition of certain components, e.g.

$$\begin{gathered} \mathcal{S}_0 = \big(b_r, a_r, \dots, b_1, a_1\big)\,, \\ \overline{\mathcal{S}_0} = \big(\overline{b_r}, \overline{a_r}, \dots, \overline{b_1}, \overline{a_1}\big)\,, \quad \mathcal{S}_0^T = \big(a_1, b_1, \dots, a_r, b_r\big)\,. \end{gathered} \tag{2.7a}$$

That is, when setting $a_1 = 0$, we impose the following configurations

$$\begin{gathered} \mathcal{S}_0 = \big(b_r, a_r, \dots, b_2, a_2, b_1\big)\,, \\ \mathcal{S}^{(\mathrm{s})} = \big(\mathcal{S}_0^T, a_{r+1}, \mathcal{S}_0, 0\big)\,, \\ a_{r+1} \in \mathbb{R}\,, \quad \mathcal{S}^{(\mathrm{sc})} = \big(\overline{\mathcal{S}_0}^T, a_{r+1}, \mathcal{S}_0, 0\big)\,, \\ a_{r+1} \in \mathbb{R}\,, \quad \mathcal{S}_1 = \big(\overline{\mathcal{S}_0}^T, a_{r+1}, \mathcal{S}_0\big)\,, \quad \mathcal{S}^{(\mathrm{ac})} = \big(\overline{\mathcal{S}_1}, \mathcal{S}_1, 0\big)\,. \end{gathered} \tag{2.7b}$$

For $a_1 \neq 0$, we instead impose the configurations

$$\begin{gathered} \mathcal{S}_0 = \big(a_r, \dots, b_2, a_2, b_1, a_1\big)\,, \\ \mathcal{S}^{(\mathrm{s})} = \big(0, \mathcal{S}_0^T, b_r, \mathcal{S}_0\big)\,, \\ b_r \in \mathbb{R}\,, \quad \mathcal{S}^{(\mathrm{sc})} = \big(0, \overline{\mathcal{S}_0}^T, b_r, \mathcal{S}_0\big)\,, \\ b_r \in \mathbb{R}\,, \quad \mathcal{S}_1 = \big(\overline{\mathcal{S}_0}^T, b_r, \mathcal{S}_0\big)\,, \quad \mathcal{S}^{(\mathrm{ac})} = \big(0, \overline{\mathcal{S}_1}, \mathcal{S}_1\big)\,. \end{gathered} \tag{2.7c}$$

Further information is found in [33, 34, 36, 37].

Examples. Widely used low-order exponential operator splitting methods such as the non-symmetric first-order Lie–Trotter splitting method and the symmetric second-order Strang splitting method are retained for the particular choices

$$\begin{aligned}&\text{Lie–Trotter}: \quad p = s = 1\,, \quad \mathcal{S} = (1,1)\,,\\&\text{Strang}: \quad p = s = 2\,, \quad \mathcal{S} = \left(\tfrac{1}{2}, 1, \tfrac{1}{2}, 0\right),\end{aligned}$$

see (2.4), (2.5), and (2.6). The method coefficients of various higher-order exponential operator splitting methods are collected in a MATLAB code that is available at `doi.org/10.5281/zenodo.13834638`, see also Section 5. Their additional features and stability properties are displayed in Table 1.

Adaptivity. For non-reversible systems such as reaction-diffusion equations and associated linear equations, see (3.3) below, it is desirable to enhance the efficiency of long-term simulations by incorporating a time stepsize control. Further details are found in [36, 44], where two different strategies have been exploited, see also the references given therein.

3. Model problems

In this section, we first introduce our general model problem and then state high-order reaction-diffusion equations and complex Ginzburg–Landau-type equations that are included as special cases. Fundamental questions such as well-posedness are treated in [27–30].

General model problem. For a space-time-dependent complex-valued solution $U : \Omega \times [t_0, T] \subset \mathbb{R}^d \times \mathbb{R} \to \mathbb{C}$, we consider the model problem

$$\begin{cases}\partial_t U(x,t) = \displaystyle\sum_{k=0}^{K} \alpha_k\, \Delta^k\, U(x,t) + W(x)\, U(x,t) + f\big(U(x,t)\big)\,,\\ U(x,t_0) = U_0(x)\,, \quad (x,t) \in \Omega \times [t_0, T]\,,\end{cases} \tag{3.1}$$

which comprises a linear combination of powers of the Laplace operator, a sufficiently regular space-dependent function $W : \Omega \to \mathbb{C}$, and a nonlinear multiplication operator defined by the complex function $f : \mathbb{C} \to \mathbb{C}$. Throughout, we assume that the complex constants $(\alpha_k)_{k=0}^{K} \in \mathbb{C}^{K+1}$ are chosen such that well-posedness of the problem is ensured. The term for $k = 0$ is insignificant and could instead be included in W or f.

Compact reformulation. In order to rewrite (3.1) in the compact form (2.1), we set $u(t) = U(\cdot, t)$ for $t \in [t_0, T]$ and define the associated operators for any sufficiently regular space-dependent function $v : \Omega \to \mathbb{C}$ through

$$\begin{aligned} \big(F_1(v)\big)(x) &= \sum_{k=0}^{K} \alpha_k \, \Delta^k \, v(x)\,, \\ \big(F_2(v)\big)(x) &= W(x)\, v(x) + f\big(v(x)\big)\,. \end{aligned} \tag{3.2}$$

3.1. *High-order reaction-diffusion equations*

We study high-order reaction-diffusion equations of the form

$$\begin{cases} \partial_t U(x,t) = \displaystyle\sum_{k=0}^{4} \alpha_k \, \Delta^k \, U(x,t) + \sum_{k=1}^{3} \beta_k \, \big(U(x,t)\big)^k\,, \\ U(x,t_0) = U_0(x)\,, \quad (x,t) \in \Omega \times [t_0, T]\,, \end{cases} \tag{3.3}$$

retained from (3.1) for $W = 0$ and a cubic polynomial

$$f : \mathbb{R} \longrightarrow \mathbb{R} : v \longmapsto \sum_{k=1}^{3} \beta_k \, v^k\,.$$

With regard to the simulation of quasicrystalline patterns in two space dimensions, we assume that the underlying spatial domains $\Omega \subset \mathbb{R}^d$ are given by Cartesian products of bounded intervals, tacitly impose periodic boundary conditions, and focus on real constants $(\alpha_k)_{k=0}^4 \in \mathbb{R}^5$ and $(\beta_k)_{k=1}^3 \in \mathbb{R}^3$. Under the basic assumption $\alpha_4 < 0$, the problem is well-posed, and the solution $U : \Omega \times [t_0, T] \to \mathbb{R}$ inherits the regularity of the prescribed initial state.

3.2. *Complex Ginzburg–Landau-type equations*

A particularity of Ginzburg–Landau-type equations

$$\begin{cases} \partial_t U(x,t) = \alpha_1 \, \Delta U(x,t) + \alpha_0 \, U(x,t) \\ \qquad\qquad + \beta_1 V(x)\, U(x,t) + \beta_2 \, \big|U(x,t)\big|^{\beta_3} \, \big(U(x,t)\big)^{\beta_4}\,, \\ U(x,t_0) = U_0(x)\,, \quad (x,t) \in \Omega \times [t_0, T]\,, \end{cases} \tag{3.4}$$

is the interaction of the Laplace operator and the non-holomorphic nonlinearity in the presence of complex constants. Evidently, the model problem (3.4) can be cast into the general form (3.1) when setting $\alpha_k = 0$ for $k \in \{2,3,4\}$, $W = \beta_1 V$, and

$$f : \mathbb{C} \longrightarrow \mathbb{C} : v \longmapsto \beta_2 \, |v|^{\beta_3} \, v^{\beta_4}\,.$$

In view of relevant applications, we once again suppose that the underlying spatial domains $\Omega \subset \mathbb{R}^d$ are given by Cartesian products of (sufficiently large) bounded intervals and tacitly impose periodic boundary conditions. Likewise, we conjecture sufficiently regular solutions $U : \Omega \times [t_0, T] \to \mathbb{C}$ for appropriately prescribed potentials and initial states. On the one hand, to ensure that the evolution equation in (3.4) is of parabolic type, we employ the restriction $\Re(\alpha_1) > 0$. On the other hand, it is instructive to observe that nonlinear Schrödinger equations such as Gross–Pitaevskii equations are included in (3.4) in the limiting case $\Re(\alpha_1) \to 0$ and $\Im(\alpha_1) \neq 0$.

Special cases. We next specify complex Ginzburg–Landau equations, time-dependent Gross–Pitaevskii equations, and their parabolic counterparts. In all cases, we focus on the special choices $\beta_3 = 2$ and $\beta_4 = 1$.

(i) Complex Ginzburg–Landau equations

$$\begin{cases} \partial_t U(x,t) = \alpha_1 \, \Delta U(x,t) + \alpha_0 \, U(x,t) + \beta_2 \, \big|U(x,t)\big|^2 \, U(x,t) \,, \\ U(x,t_0) = U_0(x) \,, \quad (x,t) \in \Omega \times [t_0, T] \,, \end{cases} \tag{3.5}$$

are obtained from (3.4) for $\beta_1 = 0$.

(ii) Time-dependent Gross–Pitaevskii equations

$$\begin{cases} \mathrm{i} \, \partial_t U(x,t) = \alpha \, \Delta U(x,t) + \beta \, V(x) \, U(x,t) + \vartheta \, \big|U(x,t)\big|^2 \, U(x,t) \,, \\ U(x,t_0) = U_0(x) \,, \quad (x,t) \in \Omega \times [t_0, T] \,, \end{cases} \tag{3.6}$$

result from (3.4) for the choices $\alpha_1 = -\,\mathrm{i}\,\alpha$, $\alpha_0 = 0$, $\beta_1 = -\,\mathrm{i}\,\beta$, and $\beta_2 = -\,\mathrm{i}\,\vartheta$ with real constants $\alpha, \beta, \vartheta \in \mathbb{R}$.

(iii) Related nonlinear parabolic equations

$$\begin{cases} \partial_t U(x,t) = \alpha \, \Delta U(x,t) + \beta \, V(x) \, U(x,t) + \vartheta \, \big|U(x,t)\big|^2 \, U(x,t) \,, \\ U(x,t_0) = U_0(x) \,, \quad (x,t) \in \Omega \times [t_0, T] \,, \end{cases} \tag{3.7}$$

can be cast into (3.4) with $\alpha_1 = \alpha \in \mathbb{R}_{>0}$, $\alpha_0 = 0$, $\beta_1 = \beta \in \mathbb{R}$, and $\beta_2 = \vartheta \in \mathbb{R}$.

4. Convergence analysis

Guideline. Our main objective in this section is the statement of a convergence result for high-order exponential operator splitting methods involving complex coefficients (2.4) applied to the general model problem (3.1). As special cases, high-order reaction-diffusion equations (3.3), complex Ginzburg–Landau-type equations (3.4), and reductions to related linear equations are included in the analysis. For this purpose, we reconsider the approach exploited in [17, 18] for standard real splitting methods applied to linear and nonlinear evolution equations of Schrödinger type and indicate suitable modifications. In Section 4.1, we examine potential stability issues, and in Section 4.2, we outline the derivation of local error expansions such that iterated commutators characterise the regularity requirements. For the statement of Theorem 1 within the context of abstract evolution equations (2.1), we employ the frameworks of analytic semigroups or strongly continuous unitary groups, respectively. In view of practical implementations, the treatment of full discretisations based on time-splitting methods combined with spectral space discretisations relies on more concrete settings of separable Hilbert spaces. Numerical experiments that confirm the validity of stability conditions and global errors bounds are presented in Section 5. They in particular imply that complex exponential operator splitting methods retain their classical orders for high-order reaction-diffusion equations and complex Ginzburg–Landau-type equations.

Fundamental means. We here refrain from a survey of the theoretical foundations and instead refer to the standard monographs [27–30]. Brief descriptions of the powerful theory of sectorial operators generating analytic semigroups on Banach spaces and applications to nonautonomous as well as quasilinear parabolic problems are also found in our former works [45, 46].

4.1. *Stability conditions*

Derivation. With regard to a stability analysis of complex exponential operator splitting methods (2.4) applied to the general model problem (3.1), we study the decisive contribution involving the highest power of the Laplace operator and a complex constant

$$\begin{cases} v'(t) = A_K\, v(t)\,, \quad A_K = \alpha_K\, \Delta^K\,, \\ v(t_n) \text{ given}\,, \quad t \in [t_n, t_n + \tau_n]\,, \quad \tau_n > 0\,. \end{cases}$$

We recall our assumption of a spatial domain that is defined by the Cartesian product of bounded intervals and the imposed periodic boundary conditions. Without loss of generality, we may consider $\Omega = [-\pi, \pi]^d$. A natural choice for the underlying function space is provided by the

Lebesgue space of complex-valued square-integrable functions

$$\left(L^2(\Omega,\mathbb{C}), (\cdot|\cdot)_{L^2}, \|\cdot\|_{L^2}\right),$$

endowed with the standard inner product and the induced norm. A complete orthonormal system of this Hilbert space is given by the family of Fourier functions $(\mathcal{F}_m)_{m\in\mathbb{Z}^d}$, which forms a set of eigenfunctions such that

$$\begin{gathered}\mathcal{F}_m : \mathbb{C}^d \longrightarrow \mathbb{C} : x = (x_1, \dots, x_d) \longmapsto \tfrac{1}{(2\,\pi)^d} \prod_{\ell=1}^{d} \mathrm{e}^{\,\mathrm{i}\, m_\ell\,(x_\ell+\pi)}\,,\\ \Delta^K \mathcal{F}_m = \big(-\lambda_m\big)^K \mathcal{F}_m\,, \quad \lambda_m = |m|^2 = \sum_{\ell=1}^{d} m_\ell^2 \geq 0\,,\\ m = (m_1, \dots, m_d) \in \mathbb{Z}^d\,.\end{gathered} \tag{4.1}$$

Due to the fact that $\lambda_m \to \infty$ whenever a component of the multi-index $m \in \mathbb{Z}^d$ tends to infinity, the term $(-\lambda_m)^K$ associated with the highest power of the Laplace operator indeed dominates lower-order contributions. By Fourier series representations, we obtain the relations

$$\begin{gathered}v(t_n) = \sum_{m\in\mathbb{Z}^d} v_m(t_n)\,\mathcal{F}_m\,,\\ \mathcal{E}_{\tau_n, a_j A_K}\, v(t_n) = \sum_{m\in\mathbb{Z}^d} \mathrm{e}^{\tau_n\, a_j\, \alpha_K\, (-\lambda_m)^K}\, v_m(t_n)\,\mathcal{F}_m\,, \quad j \in \{1, \dots, s\}\,.\end{gathered}$$

Moreover, Parseval's identity implies

$$\begin{aligned}\big\|\mathcal{E}_{\tau_n, a_j A_K}\, v(t_n)\big\|_{L^2}^2 &= \sum_{m\in\mathbb{Z}^d} \Big|\mathrm{e}^{\tau_n\, a_j\, \alpha_K\, (-\lambda_m)^K}\Big|^2 \big|v_m(t_n)\big|^2\,,\\ &= \sum_{m\in\mathbb{Z}^d} \mathrm{e}^{2\,(-1)^K\, \Re(a_j\, \alpha_K)\, \tau_n\, \lambda_m^K} \big|v_m(t_n)\big|^2\,, \quad j \in \{1, \dots, s\}\,.\end{aligned}$$

As a consequence, for positive time stepsizes $\tau_n > 0$, boundedness is ensured provided that

$$\mathrm{e}^{(-1)^K\, \Re(a_j\, \alpha_K)\, \tau_n\, \lambda_m^K} \leq 1\,, \quad m \in \mathbb{Z}^d\,.$$

Altogether, for evolution equations involving complex constants and exponential operator splitting methods with complex coefficients, we obtain the stability conditions

$$\begin{gathered}(-1)^K\, \Re\big(a_j\, \alpha_K\big) = (-1)^K \Big(\Re(a_j)\, \Re(\alpha_K) - \Im(a_j)\, \Im(\alpha_K)\Big) \leq 0\,,\\ j \in \{1, \dots, s\}\,.\end{gathered} \tag{4.2}$$

In particular, they must hold for high-order reaction-diffusion equations (3.3) with $K = 4$ and for complex Ginzburg–Landau-type equations (3.4) with $K = 1$, respectively.

Special case. It is expedient to reconsider the stability conditions (4.2) for exponential operator splitting methods involving real coefficients $(a_j)_{j=1}^{s}$. In this special case, the following simplifications are valid

$$\Im(a_j) = 0 \quad \Longrightarrow \quad (-1)^K \, \Re(a_j)\, \Re(\alpha_K) \leq 0\,, \\ j \in \{1, \dots, s\}\,.$$

On the one hand, for well-posed reaction-diffusion-type equations satisfying the relation $(-1)^K \, \Re(\alpha_K) < 0$, this yields

$$a_j = \Re(a_j) \geq 0\,, \quad j \in \{1, \dots, s\}\,.$$

We point out that the validity of these stability conditions can be granted for higher-order exponential operator splitting methods involving complex coefficients, whereas a second-order barrier is effective for any standard real splitting method. On the other hand, due to $\Re(\alpha_K) = 0$, stability can be ensured for Schrödinger equations.

Conclusion. It is noteworthy that the sizes of the real and imaginary parts of the decisive constant $\alpha_K \in \mathbb{C}$ in general affect the stability behaviour of complex exponential operator splitting methods. This conclusion from (4.2) is illustrated by a numerical experiment in Section 5 below, see Figure 5. In order to avoid such a phenomenon, it is desirable to apply high-order complex splitting methods with the additional feature $(a_j)_{j=1}^{s} \in \mathbb{R}^s$. In Table 1, we summarise the stability properties of various real and complex exponential operator splitting methods. For the stated reasons, we highlight generally applicable fourth- and sixth-order schemes with non-negative coefficients $(a_j)_{j=1}^{s}$ (Methods 5, 7, 9, 10, 11, 12).

4.2. *Local error expansions*

Fundamental means. For the purpose of illustration, we focus on the derivation of suitable local error expansions for complex operators splitting methods (2.4) applied to abstract evolution equations (2.1) that are defined by two linear operators. In view of our model problem, we associate $F_1 = A$ with a differential operator and $F_2 = B$ with a multiplication operator. Fundamental means for stepwise expansions of the exact and

Table 1. Survey of lower- and higher-order exponential operator splitting methods.

Method 1 (Lie–Trotter, real, $p = s = 1$)	Stability ($a_1 > 0$)
Method 2 (Strang, real s, $p = s = 2$)	Stability ($a_1, a_2 \geq 0$)
Method 3 (Yoshida, real s, $p = s = 4$)	Instability ($a_3 < 0$)
Method 4 (Yoshida, complex s, $p = s = 4$)	$\Re\, a_1, \dots, \Re\, a_s \geq 0$
Method 5 (complex s, $p = 4$, $s = 6$)	Stability ($a_1, \dots, a_s \geq 0$)
Method 6 (complex sc, $p = s = 3$)	$\Re\, a_1, \dots, \Re\, a_s \geq 0$
Method 7 (complex sc, $p = 3$, $s = 4$)	Stability ($a_1, \dots, a_s \geq 0$)
Method 8 (complex sc, $p = s = 4$)	$\Re\, a_1, \dots, \Re\, a_s \geq 0$
Method 9 (complex sc, $p = 4$, $s = 6$)	Stability ($a_1, \dots, a_s \geq 0$)
Method 10 (complex sc, $p = 6$, $s = 16$)	Stability ($a_1, \dots, a_s \geq 0$)
Method 11 (complex ac, $p = 4$, $s = 7$)	Stability ($a_1, \dots, a_s \geq 0$)
Method 12 (complex ac, $p = 6$, $s = 19$)	Stability ($a_1, \dots, a_s \geq 0$)

Note: First column: Specification of type (real, complex), additional features (symmetric (s), symmetric-conjugate (sc), and alternating-conjugate (ac)) and characteristics (nonstiff order p, number of stages s). Methods 1, 2, 3 involve real coefficients $(a_j, b_j)_{j=1}^s \in \mathbb{R}^{2s}$. Methods 4, 6, 8 comprise complex coefficients $(a_j, b_j)_{j=1}^s \in \mathbb{C}^{2s}$, whereas Methods 5, 7, 9, 10, 11, 12 involve real coefficients $(a_j)_{j=1}^s \in \mathbb{R}^s$ and complex coefficients $(b_j)_{j=1}^s \in \mathbb{C}^s$. Second column: Stability properties for evolution equations of parabolic type. For schemes involving complex coefficients, the conditions $\Re\, a_j \geq 0$ for $j \in \{1, \dots, s\}$ are necessary, but not sufficient. Schemes involving real coefficients $(a_j)_{j=1}^s$ remain stable for evolution equations of Schrödinger type.

numerical evolution operators, which ensure the specification of the arising remainders, are provided by the variation-of-constants formula and Taylor series expansions. The characterization of the resulting regularity requirements is linked to the identification of iterated commutators. The extension to the significantly more involved nonlinear case in the lines of [18] relies on the formal calculus of Lie-derivatives.

Linear evolution equations. We study linear evolution equations

$$u'(t) = F\big(u(t)\big) = (A + B)\, u(t)\,, \quad t \in [t_0, T]\,,$$

and employ a compact notation for the associated evolution operators

$$\mathcal{E}_{\tau,F} = \mathrm{e}^{\,\tau(A+B)}\,, \quad \mathcal{E}_{\tau,A} = \mathrm{e}^{\,\tau A}\,, \quad \mathcal{E}_{\tau,B} = \mathrm{e}^{\,\tau B}\,.$$

In order to illustrate the general procedure for high-order exponential operator splitting methods, we consider a scheme of nonstiff order p involving four stages. We meanwhile use the general form

$$s = 4: \quad \mathcal{S}_{\tau,F} = \mathrm{e}^{B_s}\,\mathrm{e}^{A_s} \cdots \mathrm{e}^{B_1}\,\mathrm{e}^{A_1}\,,$$
$$A_j = \tau\, a_j\, A\,, \quad B_j = \tau\, b_j\, B\,, \quad j \in \{1,\dots,s\}\,.$$

(i) The derivation of appropriate expansions of the exact evolution operators is based on a repeated application of the linear variation-of-constants formula

$$u(t) = \mathrm{e}^{(t-t_0)A}\, u(t_0) + \int_{t_0}^{t} \mathrm{e}^{(t-\zeta)A}\, B\, u(\zeta)\, \mathrm{d}\zeta\,, \quad t \in [t_0, T]\,.$$

Under suitable regularity restrictions, the resulting representations

$$\mathcal{E}_{\tau,F} = \mathrm{e}^{\tau(A+B)} = \sum_{k=0}^{p} \mathcal{E}_{\tau,F}^{(k)} + \mathcal{O}\big(\tau^{p+1}\big)\,, \tag{4.3}$$

where the dominant terms are given by iterated integrals such as

$$\mathcal{E}_{\tau,F}^{(0)} = \mathrm{e}^{\tau A}\,,$$
$$\mathcal{E}_{\tau,F}^{(1)} = \int_0^{\tau} \mathrm{e}^{(\tau-\zeta_1)A}\, B\, \mathrm{e}^{\zeta_1 A}\, \mathrm{d}\zeta_1\,,$$
$$\mathcal{E}_{\tau,F}^{(2)} = \int_0^{\tau}\int_0^{\zeta_1} \mathrm{e}^{(\tau-\zeta_1)A}\, B\, \mathrm{e}^{(\zeta_1-\zeta_2)A}\, B\, \mathrm{e}^{\zeta_2 A}\, \mathrm{d}\zeta_2\, \mathrm{d}\zeta_1\,,$$

are well-defined. Detailed calculations and explanations, valid for the general case, are found in [17].

(ii) For the numerical evolution operators, the derivation of appropriate representations that resemble (4.3) are obtained by Taylor series expansions of the evolution operators associated with the multiplication operators. This is done step-by-step such that the remainders reflect the regularity requirements in accordance with numerical evidence. For detailed calculations and explanations, we again refer to [17]. In the case of the above stated example, this yields a relation of the form

$$\mathcal{S}_{\tau,F} = \mathrm{e}^{B_4}\,\mathrm{e}^{A_4}\,\mathrm{e}^{B_3}\,\mathrm{e}^{A_3}\,\mathrm{e}^{B_2}\,\mathrm{e}^{A_2}\,\mathrm{e}^{B_1}\,\mathrm{e}^{A_1} = \sum_{k=0}^{p} \mathcal{S}_{\tau,F}^{(k)} + \mathcal{O}\big(\tau^{p+1}\big)\,, \tag{4.4}$$

where the leading contributions are given by

$$\begin{aligned}
\mathcal{S}^{(0)}_{\tau,F} &= \mathrm{e}^{A_4+A_3+A_2+A_1}\,,\\
\mathcal{S}^{(1)}_{\tau,F} &= B_4\,\mathrm{e}^{A_4+A_3+A_2+A_1} + \mathrm{e}^{A_4}\,B_3\,\mathrm{e}^{A_3+A_2+A_1}\\
&\quad + \mathrm{e}^{A_4+A_3}\,B_2\,\mathrm{e}^{A_2+A_1} + \mathrm{e}^{A_4+A_3+A_2}\,B_1\,\mathrm{e}^{A_1}\,,\\
\mathcal{S}^{(2)}_{\tau,F} &= \tfrac{1}{2}\,B_4^2\,\mathrm{e}^{A_4+A_3+A_2+A_1} + B_4\,\mathrm{e}^{A_4}\,B_3\,\mathrm{e}^{A_3+A_2+A_1}\\
&\quad + B_4\,\mathrm{e}^{A_4+A_3}\,B_2\,\mathrm{e}^{A_2+A_1} + B_4\,\mathrm{e}^{A_4+A_3+A_2}\,B_1\,\mathrm{e}^{A_1}\\
&\quad + \tfrac{1}{2}\,\mathrm{e}^{A_4}\,B_3^2\,\mathrm{e}^{A_3+A_2+A_1} + \mathrm{e}^{A_4}\,B_3\,\mathrm{e}^{A_3}\,B_2\,\mathrm{e}^{A_2+A_1}\\
&\quad + \mathrm{e}^{A_4}\,B_3\,\mathrm{e}^{A_3+A_2}\,B_1\,\mathrm{e}^{A_1} + \tfrac{1}{2}\,\mathrm{e}^{A_4+A_3}\,B_2^2\,\mathrm{e}^{A_2+A_1}\\
&\quad + \mathrm{e}^{A_4+A_3}\,B_2\,\mathrm{e}^{A_2}\,B_1\,\mathrm{e}^{A_1} + \tfrac{1}{2}\,\mathrm{e}^{A_4+A_3+A_2}\,B_1^2\,\mathrm{e}^{A_1}\,.
\end{aligned}$$

(iii) The order conditions for complex exponential operator splitting methods applied to linear evolution equations involving unbounded operators follow from a further analysis of the differences

$$\sum_{k=0}^{p}\Big(\mathcal{S}^{(k)}_{\tau,F} - \mathcal{E}^{(k)}_{\tau,F}\Big) = \mathcal{O}\big(\tau^{p+1}\big)\,,$$

see (4.3) and (4.4). For this purpose, the contributions in the expansions of the numerical evolution operators are understood as quadrature approximations to the integrals arising in the expansions of the exact evolution operators. Commutators naturally result from suitable Taylor series expansions, which comprise elements such as

$$\begin{aligned}
\tfrac{\mathrm{d}}{\mathrm{d}\zeta}\,\mathrm{e}^{(\tau-\zeta)A}\,B\,\mathrm{e}^{\zeta A} &= -\,\mathrm{e}^{(\tau-\zeta)A}\,\mathrm{ad}_A(B)\,\mathrm{e}^{\zeta A}\,,\\
\tfrac{\mathrm{d}^2}{\mathrm{d}\zeta^2}\,\mathrm{e}^{(\tau-\zeta)\,A}\,B\,\mathrm{e}^{\zeta A} &= \mathrm{e}^{(\tau-\zeta)A}\,\mathrm{ad}_A^{\,2}(B)\,\mathrm{e}^{\zeta A}\,,
\end{aligned}$$

see also (2.3).

Generalization to nonlinear evolution equations. The derivation of suitable local error expansions for complex exponential operator splitting methods applied to a nonlinear evolution equation of the form (2.1), retained from the general model (3.1)–(3.2), follows the approach sketched above. Important means are hereby provided by the formal calculus of Lie-derivatives. In particular, this concerns the application of the nonlinear variation-of-constants formula to obtain appropriate expansions of the exact evolution operators and the occurrence of iterated commutators involving nonlinear operators, see (2.3).

Regularity requirements. As detailed in [17], a fundamental ingredient in the rigorous treatment of high-order exponential operator splitting methods is the specification of the remainders and the characterisation of the arising commutators. In addition, the resulting regularity requirements are related to fractional power spaces of sectorial operators or, more concretely, to Sobolev spaces. For the general model problem (3.1)–(3.2), where F_1 represents a linear differential operator and F_2 a nonlinear multiplication operator, this implies that the nonstiff orders $p \in \mathbb{N}_{\geq 1}$ of the considered complex exponential operator splitting methods (2.4) are retained, provided that the problem data and hence the solution is sufficiently regular. More precisely, we can draw the following conclusions.

(i) For a linear evolution equation involving the Laplace operator and a regular space-dependent function W, straightforward calculations show that the commutator $\mathrm{ad}_{\Delta}(W)$ defines a differential operator of order one that comprises derivatives of W up to order two. By induction, it is seen that $\mathrm{ad}_{\Delta}^{p}(W)$ leads to a differential operator of order p that involves derivatives of W up to order $2\,p$. As a consequence, we retain the nonstiff order of a complex exponential operator splitting method, if the prescribed initial state and thus the exact solution is contained in the domain

$$D = D\big((-\Delta)^{p/2}\big)\,.$$

(ii) In contrast, for nonlinear problems such as complex Ginzburg–Landau-type equations, we cannot expect that certain derivatives cancel, that is, the commutator $\mathrm{ad}_{F_1}^{p}(F_2)$ corresponds to a differential operator of order $2\,p$. Hence, the nonstiff orders of convergence are retained, whenever the exact solution values remain bounded in

$$D = D\big((-\Delta)^{p}\big)\,.$$

Similar arguments are valid for high-order reaction-diffusion equations and associated linear equations.

Order reductions. In case the well-definedness of $\mathrm{ad}_{F_1}^{\ell}(F_2)$ is ensured for $\ell < p$ only, order reductions are encountered. An illustrative numerical experiment for higher-order real splitting methods applied to the Gross–Pitaevskii equation is given in [18, Figure 4.2].

4.3. *Global error estimates*

The above considerations and exemplifications lead us to the following convergence result for complex exponential operator splitting methods (2.4).

Setting. The scope of applications includes nonlinear partial differential equations of parabolic and Schrödinger type that can be cast into the general form (3.1). On the one hand, for high-order reaction-diffusion equations (3.3), complex Ginzburg–Landau equations (3.5), and parabolic counterparts of Gross–Pitaevskii equations (3.7), we employ the framework of analytical semigroups generated by sectorial operators on Banach spaces. It is noteworthy that this framework permits to include additional lower-order partial derivatives in the linear differential operators, when desired. On the other hand, for Gross–Pitaevskii equations (3.6), we instead employ the framework of strongly continuous unitary groups.

Telescopic identity. As indicated in Section 2, a standard telescopic identity reveals that estimates for the differences between numerical and exact solution values

$$\begin{aligned} u_n - u(t_n) &= \big(\mathcal{S}_{\tau_{n-1},F} \circ \cdots \circ \mathcal{S}_{\tau_0,F}\big)(u_0) - \big(\mathcal{E}_{\tau_{n-1},F} \circ \cdots \circ \mathcal{E}_{\tau_0,F}\big)\big(u(t_0)\big) \\ &= \big(\mathcal{S}_{\tau_{n-1},F} \circ \cdots \circ \mathcal{S}_{\tau_0,F}\big)(u_0) - \big(\mathcal{S}_{\tau_{n-1},F} \circ \cdots \circ \mathcal{S}_{\tau_0,F}\big)\big(u(t_0)\big) \\ &\quad + \sum_{\ell=1}^{n} \Big(\big(\mathcal{S}_{\tau_{n-1},F} \circ \cdots \circ \mathcal{S}_{\tau_\ell,F} \circ \mathcal{S}_{\tau_{\ell-1},F}\big)\big(u(t_{\ell-1})\big) \\ &\qquad - \big(\mathcal{S}_{\tau_{n-1},F} \circ \cdots \circ \mathcal{S}_{\tau_\ell,F} \circ \mathcal{E}_{\tau_{\ell-1},F}\big)\big(u(t_{\ell-1})\big)\Big), \quad n \in \{1,2,\dots,N\}, \end{aligned}$$

can be reduced to stability and local error bounds such as

$$\begin{gathered} \big\|\big(\mathcal{S}_{\tau_{n-1},F} \circ \cdots \circ \mathcal{S}_{\tau_0,F}\big)(v) - \big(\mathcal{S}_{\tau_{n-1},F} \circ \cdots \circ \mathcal{S}_{\tau_0,F}\big)(w)\big\|_X \leq C\,\|v-w\|_X, \\ \big\|\mathcal{S}_{\tau_{n-1},F}(v) - \mathcal{E}_{\tau_{n-1},F}(v)\big\|_X \leq C\,\tau_{n-1}^{p+1}, \\ n \in \{1,2,\dots,N\}, \end{gathered}$$

see also Lemmas 5 and 9 in [18]. Based on the stability analysis of Section 4.1 and the derivation of local error estimates as sketched in Section 4.2, we thus obtain the following statement. We recall the reformulation of the general model problem (3.1)–(3.2) as abstract evolution equation (2.1), where F_1 represents a linear differential operator that is given by linear combinations of powers of the Laplacian and F_2 a nonlinear multiplication operator.

Theorem 1. *Let* $(X, \|\cdot\|_X)$ *denote the underlying Banach space. Consider the nonlinear evolution equation* (2.1) *involving the (unbounded) operators*

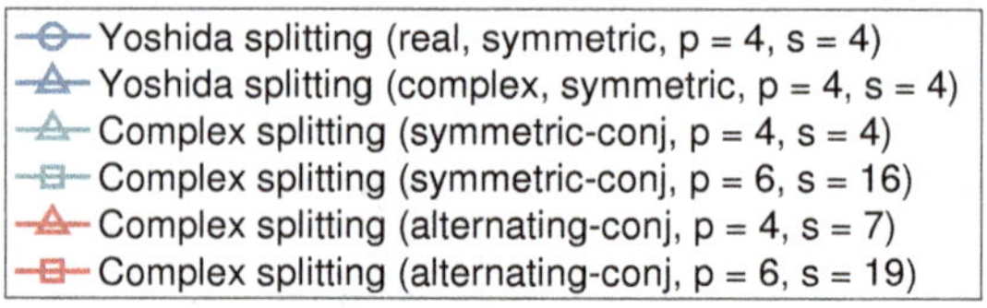

Fig. 1. Fourth- and sixth-order exponential operator splitting methods applied in numerical experiments, see Methods 3, 4, 8, 10, 11, 12 in Table 1.

$F_\ell : D(F_\ell) \subseteq X \to X$, $\ell \in \{1,2\}$. *Assume that* F_1 *generates an analytic semigroup* $(\mathcal{E}_{t,F_1})_{t\in\mathbb{R}_{\geq 0}}$ *or a strongly continuous unitary group* $(\mathcal{E}_{t,F_1})_{t\in\mathbb{R}}$, *respectively. Suppose that the coefficients of the complex exponential operator splitting method* (2.4) *fulfil the classical order conditions for some integer* $p \in \mathbb{N}_{\geq 1}$ *and that in particular the validity of the stability bounds*

$$\big\|\mathcal{E}_{t,a_j F_1}\big\|_{X\leftarrow X} \leq \mathrm{e}^{C_1 t}\,, \quad t \in [t_0, T]\,, \quad j \in \{1,\dots,s\}\,,$$

is ensured. Then, provided that the exact solution values are bounded

$$\big\|u(t)\big\|_D \leq C_2\,, \quad t \in [t_0, T]\,,$$

with respect to the norm of a suitably restricted subspace $D \subseteq X$ *that is defined by the requirement that the iterated commutators arising in the expansion of the local error remain bounded*

$$\big\|ad_{F_1}^{\ell}(F_2)\big\|_{X\leftarrow D} \leq C_3\,, \quad \ell \in \{0,1,\dots,p\}\,,$$

the following global error estimate holds

$$\big\|u_n - u(t_n)\big\|_X \leq C\left(\big\|u_0 - u(t_0)\big\|_X + \tau_{\max}^p\right),$$
$$n \in \{1,\dots,N\}\,, \quad \tau_{\max} = \max_{n\in\{0,1,\dots,N-1\}} \tau_n\,.$$

The positive constant $C > 0$ *depends on* $C_1, C_2, C_3 > 0$ *and the final time, but is independent of the time increment and the number of time steps.*

Full discretizations. We point out that the analog to Theorem 3 in [18] for space–time discretizations by high-order time-splitting Fourier pseudo-spectral methods holds as well. Its derivation relies on the setting of self-adjoint operators on Hilbert spaces. Specifically, the knowledge of a complete orthonormal system of eigenfunctions associated with the Laplacian is utilized, see (4.1).

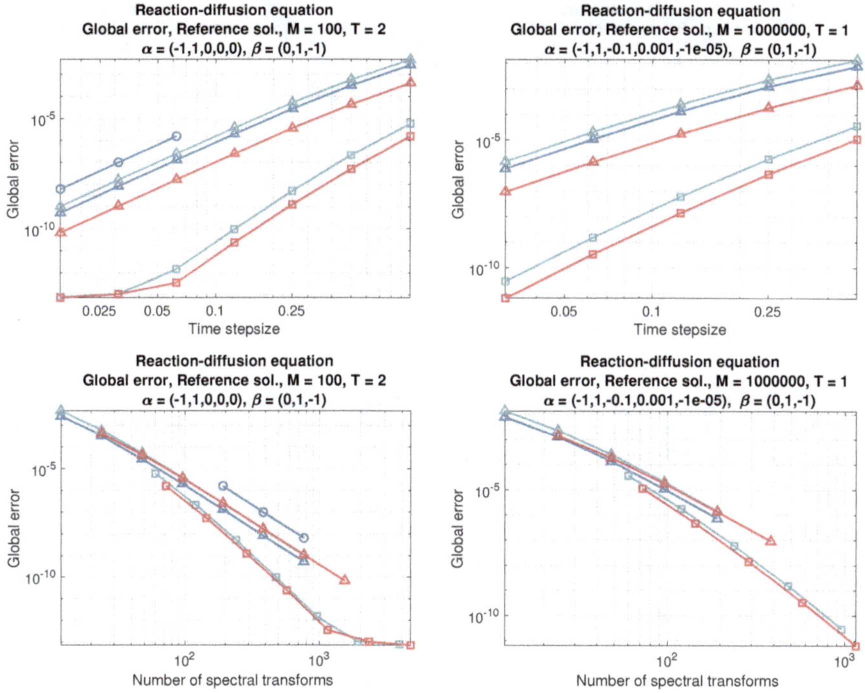

Fig. 2. Short-term integrations of a one-dimensional reaction-diffusion equation (left) and a three-dimensional high-order reaction-diffusion equation (right) by real and complex fourth- and sixth-order exponential operator splitting methods, see Figure 1. Global errors versus time stepsizes (first row) and total numbers of spectral transforms (second row), respectively. The complex exponential operator splitting methods remain stable and retain their classical orders, whereas the real splitting method by Yoshida suffers from severe instabilities.

5. Numerical experiments

In the following, we demonstrate the favorable performance of higher-order complex exponential operator splitting methods in short- and long-term integrations. We in particular confirm the validity of the stability conditions (4.2) and Theorem 1, which implies that complex exponential operator splitting methods retain their classical orders for high-order reaction-diffusion equations and complex Ginzburg–Landau-type equations.

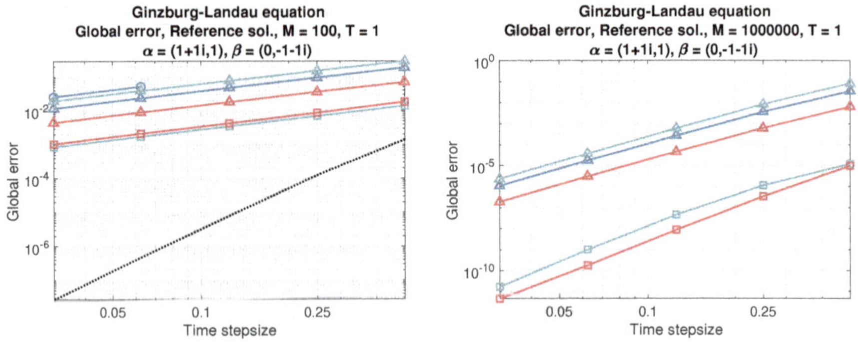

Fig. 3. Short-term integration of complex Ginzburg–Landau equations by real and complex fourth- and sixth-order exponential operator splitting methods, see Figure 1. Global errors versus time stepsizes. Left: Standard implementation for a one-dimensional problem. The complex exponential operator splitting methods remain stable, whereas the real splitting method by Yoshida becomes unstable for larger time stepsizes. The black reference line corresponds to slope four and shows that all real and complex schemes suffer from severe order reductions. Right: Correct implementation for a related three-dimensional problem. All complex exponential operator splitting methods remain stable and retain their nonstiff orders.

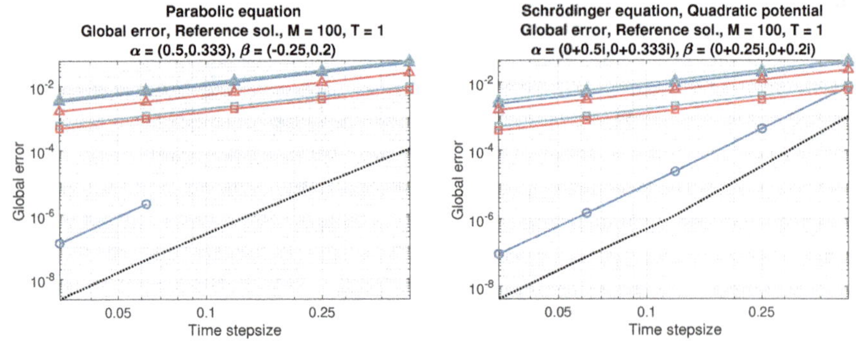

Fig. 4. Corresponding results for related equations of parabolic and Schrödinger type. For a standard implementation of higher-order complex exponential operator splitting methods, severe order reductions are observed.

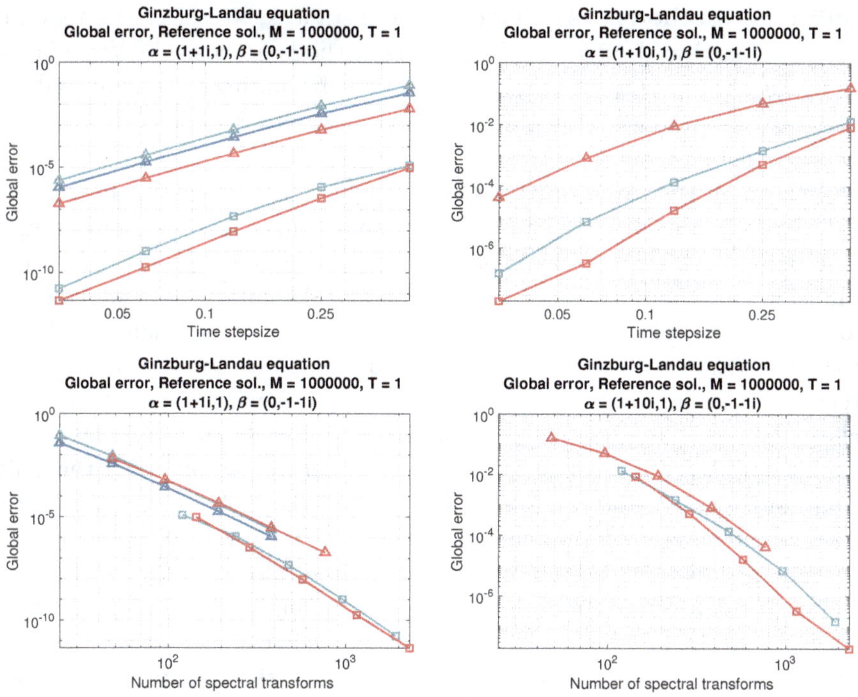

Fig. 5. Short-term integrations of three-dimensional complex Ginzburg–Landau equations by real and complex fourth- and sixth-order exponential operator splitting methods, see Figure 1. Global errors versus time stepsizes (first row) and total numbers of spectral transforms (second row), respectively. For $\alpha_1 = 1 + \mathrm{i}$, all complex exponential operator splitting methods remain stable and retain their classical orders. For $\alpha_1 = 1 + 10\,\mathrm{i}$, the stability conditions (4.2) have an impact on the schemes with complex coefficients $(a_j)_{j=1}^s$, whereas the schemes involving non-negative coefficients $(a_j)_{j=1}^s$ maintain stability.

Source. A MATLAB code that illustrates the implementation of exponential operator splitting methods for high-order reaction-diffusion equations and in particular reproduces the results shown in Figure 2 is available at

`doi.org/10.5281/zenodo.13834638.`

The coefficients of the real and complex schemes listed in Table 1 are found at the end of the file.

Splitting methods. As characteristic instances, we select the real and complex fourth-order symmetric splitting methods proposed by Yoshida [47] as well as two symmetric-conjugate and two alternating-conjugate splitting methods, in each case a fourth-order and a sixth-order scheme, see Figure 1.

Reaction-diffusion equations. In Figure 2, we include the numerical results obtained for two test cases, a one-dimensional reaction-diffusion equation as well as a computationally more demanding high-order reaction-diffusion equation in three space dimensions. As initial state, we choose a localised and highly regular Gaussian-like function. The quantity M captures the total numbers of Fourier basis functions. We compute a numerical reference solution and the errors at the final time with respect to the Euclidean norm. On the one hand, we display the global errors versus the time stepsizes. Thus, the slopes of the lines reflect the orders of the splitting methods. On the other hand, we display the global errors versus the total numbers of spectral transforms (FFT, IFFT), which corresponds to the main costs of the computations and hence relates to the efficiency of the splitting methods. In both situations, the complex exponential operator splitting methods remain stable and retain their classical orders. In contrast, due to the presence of negative coefficients, the real splitting method by Yoshida suffers from severe instabilities and yields no output for the three-dimensional high-order reaction-diffusion equation. Altogether, we observe that the sixth-order schemes are superior in efficiency, even when lower tolerances are desirable.

Ginzburg–Landau-type equations. We point out that the realisation of higher-order exponential operator splitting methods for complex Ginzburg–Landau-type equations is a subtle issue. Due to the presence of non-analytic nonlinearities and complex constants, standard implementations would lead to significant order reductions. Instead, an appropriate reformulation of (3.4) as a system for the solution and its complex conjugate is required. Additional considerations for the associated nonlinear subproblem permit the calculation of an explicit solution representation. In Figures 3 and 4, we contrast standard implementations with severe order reductions for a complex Ginzburg–Landau equation (3.5), a Gross–Pitaevskii equation (3.6), and a related parabolic equation (3.7) to a correct implementation, where the nonstiff orders are retained. In Figure 5, we demonstrate the implications of the stability conditions (4.2) for three-dimensional com-

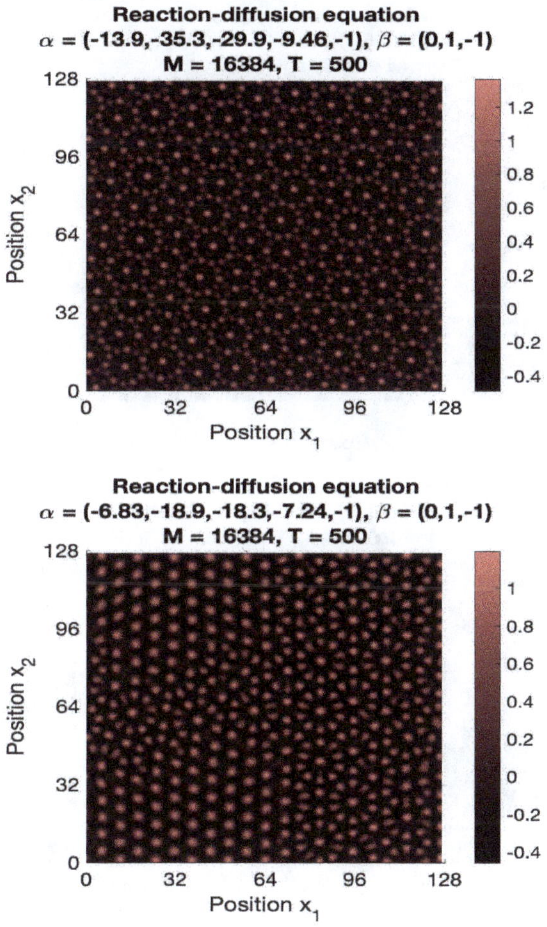

Fig. 6. Long-term integrations of two-dimensional high-order reaction-diffusion equations forming quasicrystalline patterns. Solution profiles at time $T = 500$.

plex Ginzburg–Landau equations (3.5). When increasing the imaginary part of a constant, the real splitting methods by Yoshida (Method 3) as well as the schemes involving complex coefficients $(a_j)_{j=1}^{s}$ (Methods 4 and 8) become unstable, whereas the schemes involving non-negative coefficients $(a_j)_{j=1}^{s}$ (Methods 10, 11, 12) remain stable. Overall, the sixth-order schemes are again superior in efficiency when lower tolerances are desirable.

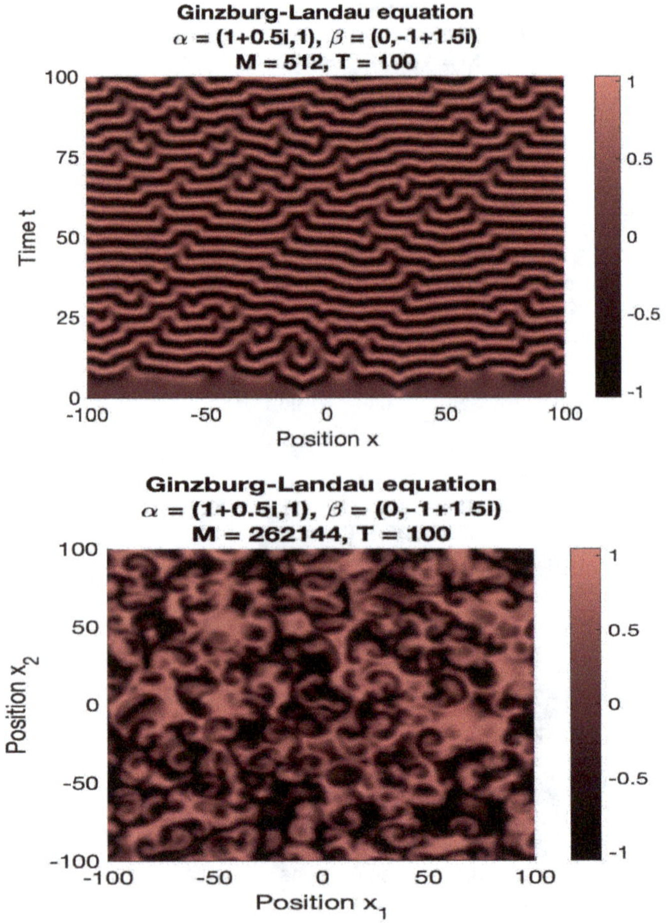

Fig. 7. Long-term integrations of one-dimensional (up) and two-dimensional (down) complex Ginzburg–Landau equations modelling nonlinear waves. Solution profiles at time $T = 100$ (real parts).

Long-term integrations. The formation of quasicrystalline patterns in long-term integrations is illustrated in Figure 6. For visualisations, see also

`techmath.uibk.ac.at/mecht/MyHomepage/Research/Movie2024Quasicrystal1.m4v`
`techmath.uibk.ac.at/mecht/MyHomepage/Research/Movie2024Quasicrystal2.m4v`

Long-term integrations of one- and two-dimensional complex Ginzburg–Landau equations modelling nonlinear waves are illustrated in Figure 7, see also

`techmath.uibk.ac.at/mecht/MyHomepage/Research/Movie2024GinzburgLandau1.m4v`
`techmath.uibk.ac.at/mecht/MyHomepage/Research/Movie2024GinzburgLandau2.m4v`

Acknowledgements

Part of this work was developed during a research stay at the Wolfgang Pauli Institute Vienna; the authors are grateful to the director Norbert Mauser and the staff members for their support and hospitality. This work has been funded by Ministerio de Ciencia e Innovación (Spain) through projects PID2022-136585NB-C21 and PID2022-136585NB-C22, MCIN/AEI/10.13039/501100011033/FEDER, UE. Sergio Blanes and Fernando Casas acknowledge the support of the Conselleria d'Innovació, Universitats, Ciència i Societat Digital from the Generalitat Valenciana (Spain) through project CIAICO/2021/180.

References

[1] E. Hairer, C. Lubich, and G. Wanner, *Geometric numerical integration*, second edn. vol. 31, *Springer Series in Computational Mathematics*, Springer-Verlag, Berlin (2006). Structure-preserving algorithms for ordinary differential equations.

[2] R. I. McLachlan and G. R. W. Quispel, Splitting methods, *Acta Numer.* **11**, 341–434 (2002).

[3] J. M. Sanz-Serna and M. P. Calvo, *Numerical Hamiltonian problems*, reprint of the 1994 original published by Chapman & Hall edn. Mineola, NY: Dover Publications (2018).

[4] S. Blanes, F. Casas, and A. Murua, Splitting methods for differential equations, *Acta Numer.* **33**, 1–161 (2024).

[5] E. P. Gross, Structure of a quantized vortex in boson systems, *Nuovo Cimento (10).* **20**, 454–477 (1961).

[6] A. Messiah. Quantum mechanics. Vol. I, II. Amsterdam: North-Holland Publishing Company, XV, p. 1-504 (1961); XV, p. 507-1136 (1962). (1962).

[7] L. Pitaevskii, Vortex lines in an imperfect Bose gas, *Sov. Phys. JETP.* **13**, 451–454 (1961).

[8] L. Pitaevskii and S. Stringari, *Bose-Einstein condensation.* vol. 116, *International Series of Monographs on Physics*, The Clarendon Press, Oxford University Press, Oxford (2003).

[9] W. Bao, Ground states and dynamics of multicomponent Bose-Einstein condensates, *Multiscale Model. Simul.* **2**(2), 210–236 (2004).

[10] W. Bao, S. Jin, and P. A. Markowich, On time-splitting spectral approximations for the Schrödinger equation in the semiclassical regime, *J. Comput. Phys.* **175**(2), 487–524 (2002).

[11] C. Besse, B. Bidégaray, and S. Descombes, Order estimates in time of splitting methods for the nonlinear Schrödinger equation, *SIAM J. Numer. Anal.* **40**(1), 26–40 (2002).

[12] S. Blanes, F. Casas, and A. Murua, An efficient algorithm based on splitting for the time integration of the Schrödinger equation, *J. Comput. Phys.* **303**, 396–412 (2015).

[13] M. Caliari and S. Zuccher, A fast time splitting finite difference approach to Gross-Pitaevskii equations, *Commun. Comput. Phys.* **29**(5), 1336–1364 (2021).

[14] R. Carles and C. Su, Scattering and uniform in time error estimates for splitting method in NLS, *Found. Comput. Math.* **24**(2), 683–722 (2024).

[15] A. Iserles and K. Kropielnicka, An elementary approach to splittings of unbounded operators, *Comput. Math. Appl.* **176**, 29–34 (2024).

[16] T. Jahnke and C. Lubich, Error bounds for exponential operator splittings, *BIT.* **40**(4), 735–744 (2000).

[17] M. Thalhammer, High-order exponential operator splitting methods for time-dependent Schrödinger equations, *SIAM J. Numer. Anal.* **46**(4), 2022–2038 (2008).

[18] M. Thalhammer, Convergence analysis of high-order time-splitting pseudospectral methods for nonlinear Schrödinger equations, *SIAM J. Numer. Anal.* **50**(6), 3231–3258 (2012).

[19] M. Thalhammer and J. Abhau, A numerical study of adaptive space and time discretisations for Gross-Pitaevskii equations, *J. Comput. Phys.* **231** (20), 6665–6681 (2012).

[20] G. Bertoli and G. Vilmart, Strang splitting method for semilinear parabolic problems with inhomogeneous boundary conditions: a correction based on the flow of the nonlinearity, *SIAM J. Sci. Comput.* **42**(3), A1913–A1934 (2020).

[21] B. Tapley, E. Celledoni, B. Owren, and H. I. Andersson, A novel approach to rigid spheroid models in viscous flows using operator splitting methods, *Numer. Algorithms.* **81**(4), 1423–1441 (2019).

[22] W. Auzinger, H. Hofstätter, and O. Koch, Non-existence of generalized splitting methods with positive coefficients of order higher than four, *Appl. Math. Lett.* **97**, 48–52 (2019).

[23] S. Blanes and F. Casas, On the necessity of negative coefficients for operator splitting schemes of order higher than two, *Appl. Numer. Math.* **54**(1), 23–37 (2005).

[24] A. Bandrauk and H. Shen, Improved exponential split operator method for solving the time-dependent Schrödinger equation, *Chem. Phys. Lett.* **176**, 428–432 (1991).

[25] F. Castella, P. Chartier, S. Descombes, and G. Vilmart, Splitting methods with complex times for parabolic equations, *BIT.* **49**(3), 487–508 (2009).

[26] E. Hansen and A. Ostermann, High order splitting methods for analytic semigroups exist, *BIT.* **49**(3), 527–542 (2009).

[27] K.-J. Engel and R. Nagel, *One-parameter semigroups for linear evolution equations.* vol. 194, *Graduate Texts in Mathematics*, Springer-Verlag, New

York (2000). With contributions by S. Brendle, M. Campiti, T. Hahn, G. Metafune, G. Nickel, D. Pallara, C. Perazzoli, A. Rhandi, S. Romanelli and R. Schnaubelt.

[28] E. Hille and R. S. Phillips, *Functional analysis and semi-groups.* American Mathematical Society Colloquium Publications, Vol. XXXI, American Mathematical Society, Providence, RI (1974). Third printing of the revised edition of 1957.

[29] A. Lunardi, *Analytic semigroups and optimal regularity in parabolic problems.* Modern Birkhäuser Classics, Birkhäuser/Springer Basel AG, Basel (1995). [2013 reprint of the 1995 original] [MR1329547].

[30] A. Pazy, *Semigroups of linear operators and applications to partial differential equations.* vol. 44, *Applied Mathematical Sciences*, Springer-Verlag, New York (1983).

[31] S. Blanes, F. Casas, P. Chartier, and A. Murua, Optimized high-order splitting methods for some classes of parabolic equations, *Math. Comp.* **82**(283), 1559–1576 (2013).

[32] W. Auzinger, H. Hofstätter, and O. Koch, Solving the Schrödinger eigenvalue problem by the imaginary time propagation technique using splitting methods with complex coefficients, *J. Chem. Phys.* **139**, 124117 (2013).

[33] J. Bernier, S. Blanes, F. Casas, and A. Escorihuela-Tomàs, Symmetric-conjugate splitting methods for linear unitary problems, *BIT.* **63**(4), Paper No. 58, 26 (2023).

[34] S. Blanes, F. Casas, P. Chartier, and A. Escorihuela-Tomàs, On symmetric-conjugate composition methods in the numerical integration of differential equations, *Math. Comp.* **91**(336), 1739–1761 (2022).

[35] S. Blanes, F. Casas, and A. Escorihuela-Tomàs, Applying splitting methods with complex coefficients to the numerical integration of unitary problems, *J. Comput. Dyn.* **9**(2), 85–101 (2022).

[36] S. Blanes, F. Casas, C. González, and M. Thalhammer, Symmetric-conjugate splitting methods for evolution equations of parabolic type, *J. Comput. Dyn.* **11**(1), 108–134 (2024).

[37] J. Bernier, S. Blanes, F. Casas, and A. Escorihuela-Tomàs. On alternating-conjugate splitting methods. arXiv preprint arXiv:2503.08453.

[38] K. Jiang and P. Zhang, Numerical methods for quasicrystals, *J. Comput. Phys.* **256**, 428–440 (2014).

[39] W. van Saarloos. The complex Ginzburg–Landau equation for beginners. In eds. P. Cladis and P. Palffy-Muhoray, *Spatio-temporal patterns in nonequilibrium complex systems*, pp. 19–32. Addison-Wesley (1995).

[40] V. S. Varadarajan, *Lie groups, Lie algebras, and their representations.* vol. 102, *Graduate Texts in Mathematics*, Springer-Verlag, New York (1984). Reprint of the 1974 edition.

[41] F. Casas, N. Crouseilles, E. Faou, and M. Mehrenberger, High-order Hamiltonian splitting for the Vlasov-Poisson equations, *Numer. Math.* **135**(3), 769–801 (2017).

[42] B. Kaltenbacher, V. Nikolic, and M. Thalhammer, Efficient time integration methods based on operator splitting and application to the Westervelt equation, *IMA J. Numer. Anal.* **35**(3), 1092–1124 (2015).

[43] Y. Li, Y. He, Y. Sun, J. Niesen, H. Qin, and J. Liu, Solving the Vlasov-Maxwell equations using Hamiltonian splitting, *J. Comput. Phys.* **396**, 381–399 (2019).

[44] S. Blanes, F. Casas, and M. Thalhammer, Splitting and composition methods with embedded error estimators, *Appl. Numer. Math.* **146**, 400–415 (2019).

[45] S. Blanes, F. Casas, and M. Thalhammer, Convergence analysis of high-order commutator-free quasi-Magnus exponential integrators for nonautonomous linear evolution equations of parabolic type, *IMA J. Numer. Anal.* **38**(2), 743–778 (2018).

[46] C. González, A. Ostermann, and M. Thalhammer, A second-order Magnus-type integrator for nonautonomous parabolic problems, *J. Comput. Appl. Math.* **189**(1-2), 142–156 (2006).

[47] H. Yoshida, Construction of higher order symplectic integrators, *Phys. Lett. A.* **150**(5-7), 262–268 (1990).

https://doi.org/10.1142/9789819816620_0004

Chapter 4

Uniform-in-Time Error Estimates for Filtered Lie Splitting Method in NLS

Rémi Carles* and Chunmei Su†

*Univ Rennes, CNRS, IRMAR-UMR 6625, F-35000 Rennes, France
Remi.Carles@math.cnrs.fr

†Yau Mathematical Sciences Center, Tsinghua University
100084 Beijing, China
sucm@tsinghua.edu.cn

We investigate the nonlinear Schrödinger equation with a defocusing nonlinearity that is mass-supercritical and energy-subcritical. Uniform-in-time first-order convergence is established for the filtered Lie-Trotter time-splitting discretization. This uniformity is achieved through the application of discrete Strichartz estimate as well as the Galilean operator, which provides precise dispersive estimates for both the exact and numerical solutions. The decay with an explicit rate in time allows us to convert traditional local error estimates, which are valid only for finite time, into a uniform-in-time error estimate applicable for all $t \in \mathbb{R}_+$.

Contents

1. Introduction

We investigate large-time error estimates for the Lie-Trotter time splitting method for the defocusing nonlinear Schrödinger equation

$$i\partial_t u + \frac{1}{2}\Delta u = |u|^{2\sigma}u, \quad (t,x) \in \mathbb{R} \times \mathbb{R}^d, \quad u_{|t=0} = \phi, \tag{1.1}$$

in space dimension $d \leq 5$, where the nonlinearity is mass-(super)critical and energy-subcritical. Specifically, this occurs when

$$\frac{2}{d} \leq \sigma < \frac{2}{(d-2)_+}, \tag{1.2}$$

that is, $\sigma \geq 2/d$ for $d \leq 2$, and $2/d \leq \sigma < 2/(d-2)$ for $d \geq 3$. The limitation on space dimension is necessary to ensure that the nonlinearity remains energy-subcritical and its derivative is locally Lipschitz continuous, thus requiring $\sigma \geq 1/2$. Under these conditions, the Cauchy problem (1.1) is globally well-posed in $H^1(\mathbb{R}^d)$ (see [1, 2]), and mass and energy are conserved,

$$\text{Mass: } M(u(t)) := \|u(t)\|^2_{L^2(\mathbb{R}^d)} = M(\phi),$$

$$\text{Energy: } E(u(t)) := \frac{1}{2}\|\nabla u(t)\|^2_{L^2(\mathbb{R}^d)} + \frac{1}{\sigma+1}\|u(t)\|^{2\sigma+2}_{L^{2\sigma+2}(\mathbb{R}^d)} = E(\phi).$$

Following [3], we denote by $\Sigma \subset H^1(\mathbb{R}^d)$ the Hilbert space, commonly referred to as the *conformal space* in the context of nonlinear Schrödinger equations,

$$\Sigma := \left\{\phi \in H^1(\mathbb{R}^d);\ \int_{\mathbb{R}^d} |x|^2|\phi(x)|^2 dx < \infty\right\},$$

which is equipped with the norm

$$\|\phi\|^2_\Sigma := \|\phi\|^2_{L^2(\mathbb{R}^d)} + \|\nabla\phi\|^2_{L^2(\mathbb{R}^d)} + \|x\phi\|^2_{L^2(\mathbb{R}^d)}.$$

Consequently, the Cauchy problem (1.1) is globally well-posed in Σ as well: if $\phi \in \Sigma$, the solution $u(t,\cdot)$ retains a finite momentum in $L^2(\mathbb{R}^d)$ for all time, and the evolution of this quantity is governed by the *pseudo-conformal evolution law*, from [3], as detailed in Proposition 5.

Next we revisit the definition of the Lie-Trotter time splitting method for (1.1). Denote $N(t)\phi$ as the solution to the nonlinear flow

$$i\partial_t u = |u|^{2\sigma}u, \quad u_{|t=0} = \phi,$$

specifically given by $N(t)\phi = \phi e^{-it|\phi|^{2\sigma}}$. We denote $S(t)\phi$ as the solution to the linear Schrödinger flow

$$i\partial_t u + \frac{1}{2}\Delta u = 0, \quad u_{|t=0} = \phi,$$

expressed as $S(t)\phi = e^{i\frac{t}{2}\Delta}\phi$. Consequently, $S(t)$ serves as a unitary operator on $H^s(\mathbb{R}^d)$ for any $s \in \mathbb{R}$. The Lie-Trotter approximation is defined, for $\tau \in (0,1)$, as

$$Z(n\tau)\phi = (S(\tau)N(\tau))^n\,\phi.$$

Error estimates for this time discretization were initially established in [4] for globally Lipschitz nonlinearities. Subsequently, Lubich [5] demonstrated error estimates for the second-order Strang splitting, accommodating the Schrödinger-Poisson nonlinearity and cubic nonlinearity (i.e., $\sigma = 1$ in (1.1)), which does not meet the criteria for global Lipschitz continuity.

The papers [4] and [5] consider data within the spaces $H^2(\mathbb{R}^d)$ for $d \leq 2$ (in [4]) and $H^4(\mathbb{R}^d)$ for $d = 3$ (in [5]), indicating that these function spaces are Banach algebras embedded into $L^\infty(\mathbb{R}^d)$. It is noteworthy that this also applies to $H^s(\mathbb{R}^d)$ when $s > d/2$. In the context of the Cauchy problem (1.1), working in $H^s(\mathbb{R}^d)$ with $s > d/2$ makes it possible to construct strong solutions locally in time; however, the existence of global solutions cannot be inferred from this approach, except when $d = s = 1$. This limitation arises because the a priori estimates derived from the conservation of mass and energy exhibit restricted regularity. The utilization of dispersive estimates marked a significant advancement in the analysis of (1.1), as detailed in [2] and later identified as Strichartz estimates in [6].

As noted in [7–10], the operator $S(\cdot)$ does not satisfy discrete-in-time Strichartz estimates. This limitation complicates the deviation of error estimates for initial data ϕ with relatively low spatial regularity. Further discussions in [11, 12] reinforce this conclusion, suggesting that a modification of $S(\cdot)$ is necessary to achieve improved convergence results. In line with the approaches outlined in [7, 9] and the adaptations proposed in [13], we introduce the modified splitting operator defined as follows:

$$Z_\tau(n\tau) = (S_\tau(\tau)N(\tau))^n \, \Pi_\tau \phi. \tag{1.3}$$

Here $S_\tau(t)$ represents the frequency-localized Schrödinger flow given by

$$S_\tau(t)\phi = S(t)\Pi_\tau\phi,$$

where

$$\widehat{\Pi_\tau \phi}(\xi) = \chi(\tau^{1/2}\xi)\widehat{\phi}(\xi), \quad \xi \in \mathbb{R}^d, \tag{1.4}$$

and χ is a sufficiently smooth cut-off function supported within the ball $B^d(0, 2)$ and satisfies $\chi \equiv 1$ on $B^d(0, 1)$.

To date, all of the previously mentioned error estimates associated with splitting methods for nonlinear Schrödinger equations have been established for bounded time intervals $[0, T]$, with constants that grow at most exponentially in T. To our knowledge, a similar situation pertains to linear Schrödinger equations with potential. On the other hand, global-in-time error estimates have been successfully proven within the framework of kinetic equations [14], where time decay estimates related to scattering phenomenon play a crucial role. Recently, the authors established analogous results for the nonlinear Schrödinger equation (1.1) by employing techniques related to the scattering theory in Σ to derive quantitative time decay estimates [15]. Specifically, we utilize the vector field $J(t) = x + it\nabla$, which is standard in scattering theory for (1.1), as it yields more precise decay estimates than the mixed $L_t^q L_x^r$-norms found in Strichartz estimates.

To revisit the primary results presented in [15], we introduce some notations. For any interval $I \subset [0, \infty)$, we define the space $\ell^q(n\tau \in I;\, L^r(\mathbb{R}^d))$, or simply $\ell^q(I;\, L^r)$ as the set of functions defined on $\tau\mathbb{Z} \cap I$ that take values in $L^r(\mathbb{R}^d)$, whose norm, given by

$$\|u\|_{\ell^q(I;\, L^r)} = \begin{cases} \left(\tau \sum\limits_{n\tau \in I} \|u(n\tau)\|^q_{L^r(\mathbb{R}^d)}\right)^{1/q}, & \text{if } 1 \leq q < \infty, \\ \sup\limits_{n\tau \in I} \|u(n\tau)\|_{L^r(\mathbb{R}^d)}, & \text{if } q = \infty, \end{cases} \tag{1.5}$$

is finite. We recall the concept of admissible pairs in the context of the Schrödinger equation. Note that we will not require endpoint Strichartz estimates, which occurs at the pair $(q, r) = (2, \frac{2d}{d-2})$ for $d \geq 3$.

Definition 1. A pair (q, r) is deemed admissible if $2 \leq r < \frac{2d}{d-2}$ (with $2 \leq r \leq \infty$ if $d = 1$, and $2 \leq r < \infty$ if $d = 2$), satisfying the condition

$$\frac{2}{q} = \delta(r) := d\left(\frac{1}{2} - \frac{1}{r}\right).$$

It is important to note that the range for q can be summarized as follows: $q \in (2, \infty]$ if $d \geq 2$, and $q \in [4, \infty]$ if $d = 1$.

Given its central role in the analysis of (1.1), throughout this chapter, we denote by (q_0, r_0) the admissible pair

$$(q_0, r_0) = \left(\frac{4\sigma + 4}{d\sigma}, 2\sigma + 2\right).$$

Notably, r_0 coincides with the product of its conjugate and $2\sigma + 1$, where $2\sigma + 1$ represents the homogeneity of the nonlinearity in the equation. This relationship plays a crucial role in estimating the nonlinear term.

In [15], we established uniform-in-time error estimates for the Lie-Trotter splitting method associated with (1.1): for $\phi \in \Sigma$, we have

$$\sup_{n \geq 0} \|Z_\tau(n\tau) - u(n\tau)\|_{L^2(\mathbb{R}^d)} \leq C\tau^{1/2}.$$

According to the findings in [12], for initial data in H^s where $s > 0$, the order $s/2$ is recognized as the natural order for the convergence rate of the Lie splitting scheme, under the CFL condition in (1.4). Modifying the CFL condition as in [12] makes it possible to improve this order of convergence. We now assume $\phi \in \Sigma \cap H^2(\mathbb{R}^d)$ and aim to give the dispersive estimates for the numerical solution Z_τ and prove the optimal convergence:

$$\sup_{n\geq 0} \|Z_\tau(n\tau) - u(n\tau)\|_{L^2(\mathbb{R}^d)} \leq C\tau.$$

Theorem 2. *Let $d \leq 5$, with σ satisfying* (1.2), *and in addition $\sigma > 1/2$ for $d = 4$ or $d = 5$. Assume that $\phi \in \Sigma \cap H^2(\mathbb{R}^d)$. Then, for any admissible pair (q, r), there exists a constant $C(d, \sigma, q, \phi)$ such that for all $\tau \in (0, 1)$, the numerical solution Z_τ satisfies*

$$\max_{A \in \{\mathbf{1}, \nabla, J, \nabla^2\}} \|A(n\tau)Z_\tau(n\tau)\|_{\ell^q(\tau\mathbb{N};L^r)} \leq C(d, \sigma, q, \phi).$$

Theorem 3. *Let $d \leq 5$, with σ satisfying* (1.2), *and where $\sigma \geq 1/2$ if $d = 5$. Consider $\phi \in \Sigma \cap H^2(\mathbb{R}^d)$ and $u \in C(\mathbb{R}; \Sigma)$ as the solution to* (1.1). *Suppose that there exists a constant M such that*

$$\max_{A \in \{\mathbf{1}, \nabla, J\}} \|A(n\tau)Z_\tau(n\tau)\|_{\ell^\infty(\tau\mathbb{N};L^2)} + \|Z_\tau(n\tau)\|_{\ell^{q_0}(\tau\mathbb{N};W^{2,r_0})} \leq M, \quad (1.6)$$

where $J(t) = x + it\nabla$. Then, there exists a constant $C = C(d, \sigma, \|\phi\|_{\Sigma\cap H^2}, M)$ such that for all $\tau \in (0, 1)$,

$$\sup_{n\geq 0} \|Z_\tau(n\tau) - u(n\tau)\|_{L^2(\mathbb{R}^d)} \leq C\tau.$$

In addition, there exists $u_+ \in \Sigma$ such that

$$\lim_{k\to\infty} \sup_{n\geq k} \|Z_\tau(n\tau) - S(n\tau)u_+\|_{L^2(\mathbb{R}^d)} \leq C\tau.$$

Compared to the assumption of the primary result in [15], the main difference lies in (1.6), where we consider the W^{2,r_0} norm in space, which involves an additional derivative compared to [15, Theorem 1.3].

The remainder of this chapter is organized as follows. In Section 2, we review standard results related to (1.1), including global-in-time estimates and present new findings, particularly Theorem 6, which offers general estimates for solutions to (1.1). Section 3 focuses on general estimates involving the numerical solution. The proof of Theorem 3 is provided in Section 4. In Appendix A.1, we recall how standard Strichartz estimates are proven, and point out the modifications needed in the discrete case.

General notations

- We denote the identity operator by $\mathbf{1}$.
- The space L^r refers to the standard space $L^r(\mathbb{R}^d)$ for $1 \leq r \leq \infty$.
- For $y \in \mathbb{R}^n$ and $n \in \mathbb{N}$, the Japanese bracket is $\langle y \rangle = (1 + |y|^2)^{1/2}$.
- The symbol C represents a generic constant, which may vary from line to line.
- To indicate the dependence of constants, we use the notation $C(d, \sigma)$ or $C_{d,\sigma}$ to signify that C does not depend on other parameters, such as τ or ϕ.
- For $a, b \geq 0$, we write $a \lesssim b$ to indicate that there exists a constant C, independent of $\tau \in (0, 1)$ and the time interval considered (but certainly depending on d and σ), such that $a \leq Cb$.

2. Preliminary estimates: The exact solution

In this section, we present preliminary results related to the exact solution.

For $t_0 \in \mathbb{R}$, the Duhamel formula for the solution u to (1.1) is expressed as follows:

$$u(t) = S(t - t_0)u(t_0) - i \int_{t_0}^{t} S(t - s) \left(|u|^{2\sigma} u\right)(s)ds. \tag{2.1}$$

The standard Strichartz inequalities associated with the Schrödinger equation (see, e.g., [1, 16]) are summarized below (see Appendix A.1 for details).

Proposition 4 (Strichartz estimates). *Let $d \geq 1$ and $S(t) = e^{i\frac{t}{2}\Delta}$.*
(1) Homogeneous estimates. *For any admissible pair (q, r), there exists a constant C_q such that*

$$\|S(t)\phi\|_{L^q(\mathbb{R};L^r)} \leq C_q \|\phi\|_{L^2}, \quad \forall \phi \in L^2.$$

(2) Inhomogeneous estimates. *Denote*

$$D(F)(t, x) = \int_0^t S(t - s)F(s, x)ds.$$

For all admissible pairs (q_1, r_1) and (q_2, r_2), there exists a constant $C = C_{q_1,q_2}$ such that for all intervals $I \ni 0$,

$$\|D(F)\|_{L^{q_1}(I;L^{r_1})} \leq C \|F\|_{L^{q_2'}\left(I;L^{r_2'}\right)}, \quad \forall F \in L^{q_2'}(I; L^{r_2'}). \tag{2.2}$$

With the Strichartz and Hölder inequalities in mind, we note that r_0 satisfies

$$\frac{1}{r_0'} = \frac{2\sigma + 1}{r_0},$$

with $2\sigma+1$ representing the homogeneity of the nonlinearity in (1.1). Correspondingly, we introduce γ defined by

$$\frac{1}{q_0'} = \frac{1}{q_0} + \frac{2\sigma}{\gamma} \Longleftrightarrow \gamma = \frac{4\sigma(\sigma+1)}{2-(d-2)\sigma}. \tag{2.3}$$

We observe that, in view of (1.2), $\gamma \in (2,\infty)$. These relations will be applied multiple times, depending on whether continuous or discrete time intervals are considered, bearing in mind that $2\sigma \geq 1$:

$$\begin{aligned} &\left\| |f|^{2\sigma-1} gh \right\|_{L^{q_0'}(I;L^{r_0'})} \leq \|f\|^{2\sigma-1}_{L^\gamma(I;L^{r_0})} \|g\|_{L^\gamma(I;L^{r_0})} \|h\|_{L^{q_0}(I;L^{r_0})}, \quad \text{or} \\ &\left\| |f|^{2\sigma-1} gh \right\|_{\ell^{q_0'}(I;L^{r_0'})} \leq \|f\|^{2\sigma-1}_{\ell^\gamma(I;L^{r_0})} \|g\|_{\ell^\gamma(I;L^{r_0})} \|h\|_{\ell^{q_0}(I;L^{r_0})}. \end{aligned} \tag{2.4}$$

Next we introduce an important proposition established in [3] regarding a vector field, which is crucial for converting the local error estimates from [7, 13] into global ones:

Proposition 5 (Pseudo-conformal conservation law). *The operator*

$$J(t) = x + it\nabla$$

satisfies the following properties:

(i) $J(t) = S(t)xS(-t)$, *indicating that* J *commutes with the linear part of* (1.1),

$$\left[J(t), i\partial_t + \frac{1}{2}\Delta\right] = 0. \tag{2.5}$$

(ii) It can be factorized as

$$J(t) = it\, e^{i\frac{|x|^2}{2t}} \nabla\left(e^{-i\frac{|x|^2}{2t}}\ \cdot\right).$$

As a consequence, J *yields weighted Gagliardo-Nirenberg inequalities. For* $2 \leq r < \frac{2d}{(d-2)_+}$ ($2 \leq r \leq \infty$ *if* $d=1$), *there exists a constant* $C(d,r)$ *depending only on* d *and* r *such that*

$$\|f\|_{L^r} \leq \frac{C(d,r)}{|t|^{\delta(r)}} \|f\|^{1-\delta(r)}_{L^2} \|J(t)f\|^{\delta(r)}_{L^2}, \quad \delta(r) := d\left(\frac{1}{2} - \frac{1}{r}\right). \tag{2.6}$$

Additionally, if $F(z) = G(|z|^2)z$ *is* C^1, *then* $J(t)$ *acts like a derivative on* $F(w)$:

$$J(t)\left(F(w)\right) = \partial_z F(w) J(t)w - \partial_{\bar{z}} F(w)\overline{J(t)w}. \tag{2.7}$$

Any solution $u \in C(\mathbb{R};\Sigma)$ *to* (1.1) *satisfies the pseudo-conformal conservation law:*

$$\frac{d}{dt}\left(\frac{1}{2}\|J(t)u\|^2_{L^2} + \frac{t^2}{\sigma+1}\|u(t)\|^{2\sigma+2}_{L^{2\sigma+2}}\right) = \frac{t}{\sigma+1}(2-d\sigma)\|u(t)\|^{2\sigma+2}_{L^{2\sigma+2}}. \tag{2.8}$$

For any interval $I \subset \mathbb{R}_+$, we introduce the norms

$$\|u\|_{X(I)} := \max_{A\in\{\mathbf{1},\nabla,J\}} \sup_{t\in I} \|Au(t)\|_{L^2}, \tag{2.9}$$

$$\|u\|_{Y(I)} = \max_{A\in\{\mathbf{1},\nabla,J\}} \sup_{(q,r)\text{ admissible}} \|Au\|_{L^q(I;L^r)}, \tag{2.10}$$

$$\|u\|_{Z(I)} := \max_{A\in\{\mathbf{1},\nabla,J,\nabla^2\}} \sup_{(q,r)\text{ admissible}} \|Au\|_{L^q(I;L^r)}. \tag{2.11}$$

The spaces $X(I), Y(I), Z(I)$ are defined in terms of the finiteness of the corresponding quantities. It is evident that

$$\|u\|_{X(I)} \leq \|u\|_{Y(I)} \leq \|u\|_{Z(I)}.$$

Notably, the assumption $\sigma \geq 2/d$ implies that, for $t \geq 0$, the right-hand side of (2.8) is nonpositive. Combined with the conservation of mass and energy, this indicates that $u \in X(\mathbb{R}_+)$. Indeed, if the initial data $\phi \in \Sigma \cap H^2$, we have the following result regarding (1.1).

Theorem 6. *Let* $1 \leq d \leq 5$ *and* $\frac{2}{d} \leq \sigma < \frac{2}{(d-2)_+}$; *additionally, if* $d = 5$, *assume* $\sigma \geq 1/2$. *If* $\phi \in \Sigma \cap H^2$ *and* u *is a solution to* (1.1), *then for all* $A \in \{\mathbf{1}, \nabla, J\}$,

$$Au \in \bigcap_{(q,r)\ admissible} L^q(\mathbb{R}_+; W^{1,r}).$$

In addition, there exists $u_\pm \in \Sigma$ *such that*

$$\|S(-t)u(t) - u_\pm\|_\Sigma \underset{t\to\pm\infty}{\longrightarrow} 0.$$

Proof. We divide the proof into several steps.
(i) At the first stage, we demonstrate that $u \in Y(\mathbb{R}_+)$. We note that $u \in X(\mathbb{R}_+)$ implies, in particular, that $u \in L^\gamma(\mathbb{R}_+; L^{r_0})$, as

$$\gamma\delta(r_0) > 1. \tag{2.12}$$

This property is indeed classical in scattering theory for (1.1) and is equivalent to

$$\sigma > \frac{2 - d + \sqrt{d^2 + 12d + 4}}{4d} =: \sigma_*,$$

with $\sigma_* < 2/d$. This parameter σ_* serves as a standard lower bound for σ to establish scattering theory in Σ, as noted in [1] (and see also [17, 18] for the case $\sigma = \sigma_*$). The standard Gagliardo-Nirenberg inequality (for bounded t) and (2.6) (for large t) yield

$$\|u(t)\|_{L^{r_0}} \lesssim \langle t\rangle^{-\delta(r_0)} \|u(t)\|_{L^2}^{1-\delta(r_0)} \left(\|\nabla u(t)\|_{L^2}^{\delta(r_0)} + \|J(t)u(t)\|_{L^2}^{\delta(r_0)}\right), \tag{2.13}$$

where $\langle t\rangle = \sqrt{1+t^2}$. In view of the Hölder inequality in time, we have

$$\|u\|_{L^\gamma(I;L^{r_0})} \leq C \left\| \langle t\rangle^{-\delta(r_0)} \right\|_{L^\gamma(I)} \|u\|_{X(I)}, \tag{2.14}$$

where C does not depend on the time interval I. For $A \in \{\mathbf{1}, \nabla, J\}$, the Strichartz estimate (Proposition 4) and Hölder inequality (2.4) then yield

$$\sup_{(q,r)\text{ admissible}} \|Au\|_{L^q(I;L^r)} \lesssim \|u(t_0)\|_\Sigma + \|u\|^{2\sigma}_{L^\gamma(I;L^{r_0})} \sup_{(q,r)\text{ admissible}} \|Au\|_{L^q(I;L^r)}.$$

By partitioning $\mathbb{R}_+$ into finitely many intervals I_j such that $\|u\|_{L^\gamma(I_j;L^{r_0})}$ is sufficiently small, we conclude that $u \in Y(\mathbb{R}_+)$.

(ii) Since $\sigma \geq 1/2$, we can differentiate (1.1) twice in space. We check that γ, defined in (2.3), satisfies $\gamma \geq q_0$, as

$$\frac{1}{q_0} - \frac{1}{\gamma} = \frac{d\sigma - 2}{4\sigma} \geq 0,$$

by recalling $d\sigma \geq 2$. Therefore, there exists $\rho \geq 2$ such that (γ, ρ) is admissible. Moreover, set

$$s := d\left(\frac{1}{\rho} - \frac{1}{r_0}\right) = d\left(\frac{1}{\rho} - \frac{1}{2} + \frac{1}{2} - \frac{1}{r_0}\right) = \frac{2}{q_0} - \frac{2}{\gamma} = \frac{d\sigma - 2}{2\sigma}.$$

We have $W^{s,\rho} \hookrightarrow L^{r_0}$, and $s \in [0,1)$, given $2/d \leq \sigma < 2/(d-2)_+$. We note that for $A = J$, the derivative ∂_j and A do not commute. However, the commutation bracket is of order zero, which does not affect the estimates. Applying $A \in \{\mathbf{1}, \nabla, J\}$ to Duhamel's formula (2.1) yields, for $t_0 \in I$,

$$\|\partial_j A u\|_{L^q(I;L^r)} \lesssim \|\partial_j A(t_0) u(t_0)\|_{L^2} + \left\|\partial_j A\left(|u|^{2\sigma} u\right)\right\|_{L^{q_0'}(I;L^{r_0'})}. \tag{2.15}$$

Now, the Hölder inequality (2.4) entails

$$\begin{aligned}
\left\|\partial_j A\left(|u|^{2\sigma} u\right)\right\|_{L^{q_0'}(I;L^{r_0'})} &\lesssim \|u\|^{2\sigma}_{L^\gamma(I;L^{r_0})} \|\partial_j A u\|_{L^{q_0}(I;L^{r_0})} \\
&\quad + \|u\|^{2\sigma-1}_{L^\gamma(I;L^{r_0})} \|\partial_j u\|_{L^\gamma(I;L^{r_0})} \|Au\|_{L^{q_0}(I;L^{r_0})} \\
&\lesssim \|u\|^{2\sigma}_{L^\gamma(I;L^{r_0})} \|\partial_j A u\|_{L^{q_0}(I;L^{r_0})} \\
&\quad + \|u\|^{2\sigma-1}_{L^\gamma(I;L^{r_0})} \|\nabla u\|_{L^\gamma(I;W^{s,\rho})} \|Au\|_{L^{q_0}(I;L^{r_0})} \\
&\lesssim \|u\|^{2\sigma}_{L^\gamma(I;L^{r_0})} \|\partial_j A u\|_{L^{q_0}(I;L^{r_0})} \\
&\quad + \|u\|^{2\sigma-1}_{L^\gamma(I;L^{r_0})} \|u\|_{L^\gamma(I;W^{2,\rho})} \|Au\|_{L^{q_0}(I;L^{r_0})}.
\end{aligned}$$

We claim that in view of part (i), we can partition $\mathbb{R}_+$ into finitely many intervals such that the last term in (2.15) can be absorbed by the left-hand side on each interval. We invoke (2.14) to control the term $\|u\|_{L^\gamma(I;L^{r_0})}$. As proven in part (i), for all $B \in \{\mathbf{1}, \nabla, J\}$, $Bu \in L^{q_0}(\mathbb{R}; L^{r_0})$. Since q_0 is finite, $\mathbb{R}_+$ can be decomposed into finitely many intervals $(I_j)_{1\leq j\leq N}$ (with I_N of the form $I_N = [M, \infty)$) on which, since $\sigma \geq 1/2$,

$$\|u\|^{2\sigma}_{L^\gamma(I;L^{r_0})} + \sup_{B\in\{\mathbf{1},\nabla,J\}} \|u\|^{2\sigma-1}_{L^\gamma(I_j;L^{r_0})}\|Bu\|_{L^{q_0}(I_j;L^{r_0})} \leq \eta,$$

with $\eta > 0$ sufficiently small so the last term in (2.15) can be absorbed by the left-hand side of (2.15) on each I_j, when we consider the supremum over $A \in \{\mathbf{1}, \nabla, J\}$ and admissible pairs (q, r), recalling that (γ, ρ) is admissible. By finite induction, we infer that for all $A \in \{\mathbf{1}, \nabla, J\}$ and all admissible pairs (q, r), $Au \in L^q(\mathbb{R}_+; W^{1,r})$.

(iii) For the proof of the last assertion, we refer to [1, Theorem 7.4.1] or [19, Theorem B]). □

3. Preliminary estimates: The numerical solution

In this section, we introduce some preliminary results concerning the numerical solution (1.3).

The discrete-in-time counterpart of the $L^q_tL^r$ norms involved in the standard Strichartz estimates is defined in (1.5). For $a, n \in \mathbb{N}$ with $a+1 \leq n$, the discrete Duhamel formula associated with Z_τ is given by

$$Z_\tau(n\tau) = S_\tau((n-a)\tau)Z_\tau(a\tau) + \tau\sum_{k=a}^{n-1} S_\tau(n\tau - k\tau)\frac{N(\tau) - \mathbf{1}}{\tau}Z_\tau(k\tau), \tag{3.1}$$

which can be compared to the formula in (2.1). We now present analogues of Proposition 4 within the discrete-in-time framework, a setting which is less classical (see also Appendix A.1).

Proposition 7 (From [7, Theorem 2.1]). *Let (q, r), (q_1, r_1) and (q_2, r_2) be any admissible pairs. Then, there exist constants $C_{d,q}$ and $C_{d,q_1,q_2} > 0$ such that*

$$\|S_\tau(\cdot)\phi\|_{\ell^q(\tau\mathbb{Z};L^r)} \leq C_{d,q}\|\phi\|_{L^2}, \tag{3.2}$$

and

$$\left\|\tau\sum_{k=-\infty}^{n-1} S_\tau\left((n-k)\tau\right)f(k\tau)\right\|_{\ell^{q_1}(\tau\mathbb{Z};L^{r_1})} \leq C_{d,q_1,q_2}\|f\|_{\ell^{q_2'}(\tau\mathbb{Z};L^{r_2'})} \tag{3.3}$$

hold for all $\phi \in L^2$ and $f \in \ell^{q_2'}(\tau\mathbb{Z}; L^{r_2'})$.

Combining Propositions 4 and 7, and Christ-Kiselev lemma yields:

Corollary 8 (From [7, Lemma 4.5]). *For any admissible pairs* (q_1, r_1) *and* (q_2, r_2)*, we have, for* $A \in \{\mathbf{1}, \nabla\}$*,*

$$\left\| A \int_{s<n\tau} S_\tau(n\tau - s) f(s) ds \right\|_{\ell^{q_1}(\tau\mathbb{Z}; L^{r_1})} \leq C_{d,q_1,q_2} \|Af\|_{L^{q_2'}(\mathbb{R}; L^{r_2'})}. \quad (3.4)$$

Regarding the cut-off operator, we have the following result, which has been proven in [7, Lemma 4.4].

Lemma 9 (From [13, Lemma 2.6]). *For any* $1 < r < \infty$ *and* $s > 0$*, we have*

$$\left\|\Pi_\tau \phi - \phi\right\|_{L^r} \leq C\tau^{s/2} \left\|(-\Delta)^{s/2}\phi\right\|_{L^r}, \quad (3.5)$$

$$\|\Pi_\tau \phi\|_{L^r} \leq C\|\phi\|_{L^r}. \quad (3.6)$$

The following version of Bernstein lemma is adapted from [13]:

Lemma 10 (From [13, Lemma 2.6]). *For any* $1 \leq q \leq r < \infty$ *and* $\phi : \mathbb{R}^d \to \mathbb{C}$*, we have*

$$\left\|\nabla(\Pi_\tau \phi)\right\|_{L^r} \leq C\tau^{-\frac{1}{2}} \|\phi\|_{L^r}, \quad (3.7)$$

and

$$\|\Pi_\tau \phi\|_{L^r} \leq C\tau^{\frac{d}{2}\left(\frac{1}{r} - \frac{1}{q}\right)} \|\phi\|_{L^q}. \quad (3.8)$$

Lemma 11. *For the nonlinear term* $\frac{N(\tau)-\mathbf{1}}{\tau} v = \frac{\exp(-i\tau|v|^{2\sigma})-1}{\tau} v$*, we have*

$$\left|\frac{N(\tau) - \mathbf{1}}{\tau} v\right| = \left|\frac{\exp(-i\tau|v|^{2\sigma}) - 1}{\tau} v\right| \leq |v|^{2\sigma+1}, \quad (3.9)$$

$$\left|\nabla\left(\frac{N(\tau) - \mathbf{1}}{\tau} v\right)\right| \leq (2\sigma + 1)|v|^{2\sigma}|\nabla v|. \quad (3.10)$$

Furthermore,

$$\left|\nabla^2\left(\frac{N(\tau) - \mathbf{1}}{\tau} v\right)\right| \lesssim |v|^{2\sigma}|\nabla^2 v| + |v|^{2\sigma-1}|\nabla v|^2 + \tau|v|^{4\sigma-1}|\nabla v|^2, \quad (3.11)$$

$$\left|\frac{N(\tau) - \mathbf{1}}{\tau} v - \frac{N(\tau) - \mathbf{1}}{\tau} w\right| \lesssim \left(|v|^{2\sigma} + |w|^{2\sigma}\right)|v - w|, \quad (3.12)$$

for all $v, w \in \mathbb{C}$.

Proof. The proofs for (3.9), (3.11) and (3.12) can be found in [13, Lemma 2.5]. It remains to show (3.11). A direct calculation gives

$$\partial_j\left(\frac{N(\tau) - \mathbf{1}}{\tau} v\right) = \frac{\exp(-i\tau|v|^{2\sigma}) - 1}{\tau}\partial_j v - 2\sigma i e^{-i\tau|v|^{2\sigma}} v|v|^{2\sigma-2}\mathrm{Re}(\overline{v}\,\partial_j v),$$

and

$$\partial_{jk}\left(\frac{N(\tau)-\mathbf{1}}{\tau}v\right) = \frac{\exp(-i\tau|v|^{2\sigma})-1}{\tau}\partial_{jk}v$$
$$- 2i\sigma e^{-i\tau|v|^{2\sigma}}|v|^{2\sigma-2}\mathrm{Re}(\overline{v}\,\partial_k v)\partial_j v$$
$$- 4\tau\sigma^2 e^{-i\tau|v|^{2\sigma}}|v|^{4\sigma-4}v\mathrm{Re}(\overline{v}\,\partial_k v)\mathrm{Re}(\overline{v}\,\partial_j v)$$
$$- i\sigma e^{-i\tau|v|^{2\sigma}}\left(2\sigma|v|^{2\sigma-2}\mathrm{Re}(\overline{v}\partial_k v)\partial_j v + |v|^{2\sigma}\partial_{jk}v\right)$$
$$- i\sigma(2\sigma-2)e^{-i\tau|v|^{2\sigma}}|v|^{2\sigma-4}\mathrm{Re}(\overline{v}\partial_k v)v^2\overline{\partial_j v}$$
$$- i\sigma e^{-i\tau|v|^{2\sigma}}|v|^{2\sigma-2}\left(2v\partial_j v\partial_k v + v^2\overline{\partial_{jk}v}\right),$$

which implies

$$\left|\nabla^2\left(\frac{N(\tau)-\mathbf{1}}{\tau}v\right)\right| \leq |v|^{2\sigma}|\nabla^2 v| + C_\sigma|v|^{2\sigma-1}|\nabla v|^2 + C_\sigma\tau|v|^{4\sigma-1}|\nabla v|^2.$$

Thus, the proof is completed. □

Lemma 12 (From [7, Lemma 4.6]). *For any admissible pairs* (q_1, r_1) *and* (q_2, r_2)*, there is a constant* $C_{d,q_1,q_2} > 0$ *such that*

$$\left\|\int_{s<n\tau} S_\tau(n\tau-s)f(s)ds - \tau\sum_{k=-\infty}^{n-1} S_\tau(n\tau-k\tau)f(k\tau)\right\|_{\ell^{q_1}(\tau\mathbb{Z};L^{r_1})} \tag{3.13}$$
$$\leq C_{d,q_1,q_2}\,\tau\|\Delta f\|_{L^{q_2'}(\mathbb{R};L^{r_2'})} + C_{d,q_1,q_2}\,\tau\|\partial_t f\|_{L^{q_2'}(\mathbb{R};L^{r_2'})}.$$

Now we are ready to give the proof of Theorem 2.

Proof of Theorem 2. In [15], it has been demonstrated that for initial data $\phi \in \Sigma$, it holds

$$\max_{A\in\{\mathbf{1},\nabla,J\}} \|A(n\tau)Z_\tau(n\tau)\|_{\ell^q(\tau\mathbb{N};L^r)} \leq M, \tag{3.14}$$

where M is a constant depending on d, σ, q, and ϕ. Therefore, it suffices to establish that

$$\|\nabla^2 Z_\tau(n\tau)\|_{l^q(\tau\mathbb{N};L^r)} \leq C(d,\sigma,q,\phi), \tag{3.15}$$

for any admissible pair (q,r).

We recall that $\gamma\delta(r_0) > 1$ (see (2.12)). Therefore, from (1.5), (3.14) and (2.6), it follows that for any $\varepsilon > 0$, we can find a *finite* number $K = K(\varepsilon)$ of time intervals $I_j = [m_j, m_{j+1})$ with $m_1 = 0$, $m_j \in \mathbb{N}$ if $j \leq K$, $m_{K+1} = \infty$, and $\tau(\varepsilon) > 0$ such that for $1 \leq j \leq K$ and $0 < \tau \leq \tau(\varepsilon)$,

$$\|\langle k\tau\rangle^{-\delta(r_0)}\|_{\ell^\gamma(\tau I_j)} + \|Z_\tau(k\tau)\|_{\ell^\gamma(\tau I_j;L^{r_0})} \leq \varepsilon, \quad \mathbb{R}_+ = \bigcup_{j=1}^K I_j.$$

On each interval I_j, the discrete Duhamel formula (3.1) can be written as

$$\begin{aligned} Z_\tau(m_j\tau + n\tau) &= S_\tau(n\tau)Z_\tau(m_j\tau) \\ &\quad + \tau\sum_{k=0}^{n-1} S_\tau(n\tau - k\tau)\frac{N(\tau)-\mathbf{1}}{\tau}Z_\tau(m_j\tau + k\tau), \end{aligned} \tag{3.16}$$

for $0 \le n \le m_{j+1} - m_j$. Note that while m_{j+1} may be infinite, m_j is always finite. By applying Proposition 7, Lemma 11 and Hölder's inequality (2.4), we get

$$\begin{aligned} \|\nabla^2 Z_\tau\|_{\ell^q(\tau I_j;L^r)} &\le \|\nabla^2 Z_\tau(m_j\tau)\|_{L^2} + C\left\|\nabla^2\left(\frac{N(\tau)-\mathbf{1}}{\tau}v\right)\right\|_{\ell^{q_0'}(\tau I_j;L^{r_0'})} \\ &\le \|\nabla^2 Z_\tau(m_j\tau)\|_{L^2} + C\|Z_\tau\|^{2\sigma}_{\ell^\gamma(\tau I_j;L^{r_0})}\|\nabla^2 Z_\tau\|_{\ell^{q_0}(\tau I_j;L^{r_0})} \\ &\quad + C\|Z_\tau\|^{2\sigma-1}_{\ell^\gamma(\tau I_j;L^{r_0})}\|\nabla Z_\tau\|_{\ell^\gamma(\tau I_j;L^{r_0})}\|\nabla Z_\tau\|_{\ell^{q_0}(\tau I_j;L^{r_0})} \\ &\quad + C\tau\||Z_\tau|^{4\sigma-1}|\nabla Z_\tau|^2\|_{\ell^{q_0'}(\tau I_j;L^{r_0'})}, \\ &\le \|\nabla^2 Z_\tau(m_j\tau)\|_{L^2} + C\varepsilon^{2\sigma}\|\nabla^2 Z_\tau\|_{\ell^{q_0}(\tau I_j;L^{r_0})} \\ &\quad + CM\varepsilon^{2\sigma-1}\|\nabla^2 Z_\tau\|_{\ell^\gamma(\tau I_j;L^\rho)} \\ &\quad + C\tau\||Z_\tau|^{4\sigma-1}|\nabla Z_\tau|^2\|_{\ell^{q_0'}(\tau I_j;L^{r_0'})}, \end{aligned} \tag{3.17}$$

where C is a generic constant which depends on d, σ, q, and we utilized the Sobolev embedding $W^{1,\rho} \hookrightarrow L^{r_0}$ with (γ, ρ) being an admissible pair defined in the proof of Theorem 6.

Using Hölder's inequality (2.4), we derive

$$\begin{aligned} &\||Z_\tau|^{4\sigma-1}|\nabla Z_\tau|^2\|_{\ell^{q_0'}(\tau I_j;L^{r_0'})} \\ &\le \|\nabla Z_\tau\|_{\ell^{q_0}(\tau I_j;L^{r_0})}\|\nabla Z_\tau\|_{\ell^\gamma(\tau I_j;L^{r_0})}\|Z_\tau\|^{4\sigma-1}_{\ell^{\frac{4\sigma-1}{2\sigma-1}\gamma}(\tau I_j;L^{\frac{4\sigma-1}{2\sigma-1}r_0})} \\ &\le M\|\nabla^2 Z_\tau\|_{\ell^\gamma(\tau I_j;L^\rho)}\|Z_\tau\|^{4\sigma-1}_{\ell^{\frac{4\sigma-1}{2\sigma-1}\gamma}(\tau I_j;L^{\frac{4\sigma-1}{2\sigma-1}r_0})}. \end{aligned}$$

In light of Bernstein's inequality (3.8), we find that

$$\|Z_\tau(m_j\tau + k\tau)\|_{L^{\frac{4\sigma-1}{2\sigma-1}r_0}} \lesssim \tau^{\frac{d}{2}\left(\frac{2\sigma-1}{(4\sigma-1)r_0} - \frac{1}{r_0}\right)}\|Z_\tau(m_j\tau + k\tau)\|_{L^{r_0}},$$

where

$$\frac{d}{2}\left(\frac{2\sigma-1}{(4\sigma-1)r_0} - \frac{1}{r_0}\right) = -\frac{d}{2}\frac{2\sigma}{(4\sigma-1)r_0} = \frac{-d\sigma}{(4\sigma-1)(2\sigma+2)} = \frac{-2}{(4\sigma-1)q_0},$$

leading to

$$\begin{aligned} \|Z_\tau\|^{4\sigma-1}_{\ell^{\frac{4\sigma-1}{2\sigma-1}\gamma}(\tau I_j;L^{\frac{4\sigma-1}{2\sigma-1}r_0})} &\lesssim \tau^{-2/q_0}\|Z_\tau\|^{4\sigma-1}_{\ell^{\frac{4\sigma-1}{2\sigma-1}\gamma}(\tau I_j;L^{r_0})} \\ &\lesssim \tau^{-2/q_0}\|\langle n\tau\rangle^{-\delta(r_0)}\|^{4\sigma-1}_{\ell^{\frac{4\sigma-1}{2\sigma-1}\gamma}(\tau I_j)} M^{4\sigma-1} \\ &\lesssim \tau^{-2/q_0}\|\langle n\tau\rangle^{-\delta(r_0)}\|^{4\sigma-1}_{\ell^\gamma(\tau I_j)} M^{4\sigma-1} \\ &\le C\varepsilon^{4\sigma-1}\tau^{-2/q_0}M^{4\sigma-1}. \end{aligned} \tag{3.18}$$

Here we have employed (2.13) for the second inequality and $\frac{4\sigma-1}{2\sigma-1} > 1$ for the third one. Substituting (3.18) into (3.17), we arrive at

$$\begin{aligned}\|\nabla^2 Z_\tau\|_{\ell^q(\tau I_j;L^r)} \leq \|\nabla^2 Z_\tau(m_j\tau)\|_{L^2} + C\varepsilon^{2\sigma}\|\nabla^2 Z_\tau\|_{\ell^{q_0}(\tau I_j;L^{r_0})} \\ + CM\varepsilon^{2\sigma-1}\|\nabla^2 Z_\tau\|_{\ell^\gamma(\tau I_j;L^\rho)} \\ + CM^{4\sigma}\tau^{1-2/q_0}\varepsilon^{4\sigma-1}\|\nabla^2 Z_\tau\|_{\ell^\gamma(\tau I_j;L^\rho)}.\end{aligned} \tag{3.19}$$

Noticing that $q_0 > 2$ and $\tau \leq 1$, we can choose ε sufficiently small such that

$$C\varepsilon^{2\sigma} + CM\varepsilon^{2\sigma-1} + CM^{4\sigma}\varepsilon^{4\sigma-1} \leq 1/2,$$

which leads to

$$\begin{aligned}\max_{(q,r)\text{ admissible}} \|\nabla^2 Z_\tau\|_{\ell^q(\tau I_j;L^r)} &\leq 2\|\nabla^2 Z_\tau(m_j\tau)\|_{L^2} \leq 2\|\nabla^2 Z_\tau\|_{\ell^\infty(\tau I_{j-1};L^2)} \\ &\leq 2 \max_{(q,r)\text{ admissible}} \|\nabla^2 Z_\tau\|_{\ell^q(\tau I_{j-1};L^r)} \\ &\leq 2^2\|\nabla^2 Z_\tau(m_{j-1}\tau)\|_{L^2} \leq (\ldots) \\ &\leq 2^j\|\nabla^2 Z_\tau(m_1\tau)\|_{L^2} = 2^j\|\nabla^2 \Pi_\tau\phi\|_{L^2} \\ &\leq C2^j\|\phi\|_{H^2},\end{aligned}$$

thus the proof is completed by recalling that the number of intervals is finite and $\phi \in H^2$. □

4. Convergence

In this section, we give the proof of Theorem 3. Before that, we establish the following lemma.

Lemma 13. *For $u \in Z(\mathbb{R}_+)$, $\tau \in (0,1)$, denote*

$$\begin{aligned}\mathcal{A}(u)(n\tau) = \tau\sum_{k=0}^{n-1} S_\tau(n\tau - k\tau)\frac{N(\tau)-1}{\tau}\Pi_\tau u(k\tau) \\ + i\int_0^{n\tau} S_\tau(n\tau - s)|u|^{2\sigma}u(s)ds.\end{aligned}$$

Then for all admissible pairs (q,r), it holds

$$\|\mathcal{A}(u)\|_{\ell^q(\mathbb{N};L^r)} \lesssim \tau\left(\|u\|_{Z(\mathbb{R}_+)}^{2\sigma+1} + \|u\|_{Z(\mathbb{R}_+)}^{4\sigma+1}\right).$$

Proof. In view of (2.14), the assumptions of Lemma 13 imply that

$$u \in L^\gamma(\mathbb{R}_+;L^{r_0}),$$

where γ is given by (2.3), and $\gamma\delta(r_0) > 1$. We decompose $\mathcal{A}(u)(n\tau)$ as

$$\mathcal{A}(u)(n\tau) = \mathcal{A}_1(u)(n\tau) + \mathcal{A}_2(u)(n\tau),$$

where

$$\mathcal{A}_1(u)(n\tau) = \tau \sum_{k=0}^{n-1} S_\tau(n\tau - k\tau)\mathcal{B}_1(u)(k\tau) - \int_0^{n\tau} S_\tau(n\tau - s)\mathcal{B}_1(u)(s)ds,$$

$$\mathcal{A}_2(u)(n\tau) = \int_0^{n\tau} S_\tau(n\tau - s)\mathcal{B}_1(u)(s)ds + i\int_0^{n\tau} S_\tau(n\tau - s)|u|^{2\sigma}u(s)ds,$$

with

$$\mathcal{B}_1(u)(s) := \frac{N(\tau) - \mathbf{1}}{\tau}\Pi_\tau u(s).$$

By applying Lemma 12, we can estimate

$$\|\mathcal{A}_1(u)\|_{\ell^q(\tau\mathbb{N};L^r)} \lesssim \tau \|\Delta\mathcal{B}_1(u)\|_{L^{q_0'}(\mathbb{R}_+;L^{r_0'})} + \tau \|\partial_t\mathcal{B}_1(u)\|_{L^{q_0'}(\mathbb{R}_+;L^{r_0'})}, \tag{4.1}$$

for any admissible pair (q, r). As stated in Lemma 11, we find that

$$\begin{aligned}
&\|\Delta\mathcal{B}_1(u)\|_{L^{q_0'}(\mathbb{R}_+;L^{r_0'})} \\
&\lesssim \left\||\Pi_\tau u|^{2\sigma}\Delta\Pi_\tau u\right\|_{L^{q_0'}(\mathbb{R}_+;L^{r_0'})} + \left\||\Pi_\tau u|^{2\sigma-1}|\nabla\Pi_\tau u|^2\right\|_{L^{q_0'}(\mathbb{R}_+;L^{r_0'})} \\
&\quad + \tau \left\||\Pi_\tau u|^{4\sigma-1}|\nabla\Pi_\tau u|^2\right\|_{L^{q_0'}(\mathbb{R}_+;L^{r_0'})} \\
&\lesssim \|\Pi_\tau u\|^{2\sigma}_{L^\gamma(\mathbb{R}_+;L^{r_0})}\|\Delta\Pi_\tau u\|_{L^{q_0}(\mathbb{R}_+;L^{r_0})} \\
&\quad + \|\Pi_\tau u\|^{2\sigma-1}_{L^\gamma(\mathbb{R}_+;L^{r_0})}\|\nabla\Pi_\tau u\|_{L^\gamma(\mathbb{R}_+;L^{r_0})}\|\nabla\Pi_\tau u\|_{L^{q_0}(\mathbb{R}_+;L^{r_0})} \\
&\quad + \tau \left\||\Pi_\tau u|^{4\sigma-1}|\nabla\Pi_\tau u|^2\right\|_{L^{q_0'}(\mathbb{R}_+;L^{r_0'})} \\
&\lesssim \|u\|^{2\sigma+1}_{Z(\mathbb{R}_+)} + \tau \left\||\Pi_\tau u|^{4\sigma-1}|\nabla\Pi_\tau u|^2\right\|_{L^{q_0'}(\mathbb{R}_+;L^{r_0'})},
\end{aligned} \tag{4.2}$$

where we have utilized Hölder's inequality (2.4), Lemma 10 and the fact that $W^{1,\rho} \hookrightarrow L^{r_0}$, with (γ, ρ) being an admissible pair. It remains to estimate the last term. By applying Hölder's inequality, we get

$$\begin{aligned}
&\left\||\Pi_\tau u|^{4\sigma-1}|\nabla\Pi_\tau u|^2\right\|_{L^{q_0'}(\mathbb{R}_+;L^{r_0'})} \\
&\le \|\Pi_\tau u\|^{4\sigma-1}_{L^{q_2}(\mathbb{R}_+;L^{(4\sigma+1)r_0'})} \|\nabla\Pi_\tau u\|^2_{L^{q_2}(\mathbb{R}_+;L^{(4\sigma+1)r_0'})},
\end{aligned}$$

where $q_2 := (4\sigma + 1)q_0'$. By Lemma 10, we have

$$\|\Pi_\tau u\|_{L^{(4\sigma+1)r_0'}} \le C\tau^{\frac{d}{2}\left(\frac{1}{(4\sigma+1)r_0'} - \frac{1}{r_1}\right)}\|\Pi_\tau u\|_{L^{r_1}},$$

for all $r_1 \le (4\sigma + 1)r_0'$. This implies

$$\begin{aligned}
&\left\||\Pi_\tau u|^{4\sigma-1}|\nabla\Pi_\tau u|^2\right\|_{L^{q_0'}(\mathbb{R}_+;L^{r_0'})} \\
&\lesssim \tau^{\frac{(4\sigma+1)d}{2}\left(\frac{1}{(4\sigma+1)r_0'} - \frac{1}{r_1}\right)}\|\Pi_\tau u\|^{4\sigma-1}_{L^{q_2}(\mathbb{R}_+;L^{r_1})} \|\nabla\Pi_\tau u\|^2_{L^{q_2}(\mathbb{R}_+;L^{r_1})}.
\end{aligned} \tag{4.3}$$

We determine the value of $r_1 > 0$ by

$$\frac{1}{r_1} = \frac{1}{4\sigma+1}\left(\frac{2\sigma+1}{2\sigma+2} + \frac{1}{d}\right) = \frac{1}{(4\sigma+1)r_0'} + \frac{1}{(4\sigma+1)d}.$$

We claim that $r_1 \geq 2$. Indeed, by definition, $r_1 \geq 2$ if and only if

$$d\sigma(4\sigma+3) \geq 2\sigma+2,$$

which is clearly true since $d\sigma \geq 2$. On the other hand, we observe that

$$\frac{1}{q_2} = \frac{1}{(4\sigma+1)q_0'} = \frac{4(\sigma+1)-d\sigma}{4(\sigma+1)(4\sigma+1)},$$

which implies $1/q_2 > 0$ for $d \leq 5$, and $1/q_2 < 1/2$ since $\sigma \geq 2/d$. Thus, $q_2 \in (2,\infty)$, and if $d = 1$, then $q_2 > 4$. This allows us to define r_2 by

$$\frac{1}{r_2} := \frac{1}{2} - \frac{2}{dq_2},$$

such that (q_2, r_2) is admissible. The Sobolev embedding states that

$$W^{s,r_2} \hookrightarrow L^{r_1}, \quad \frac{1}{r_1} + \frac{s}{d} = \frac{1}{r_2} \Longleftrightarrow s = \frac{d}{2} - \frac{d+6}{8\sigma+2}.$$

We note that for $\frac{2}{d} \leq \sigma < \frac{2}{(d-2)_+}$, it holds that $s \in (0,1)$, hence (4.3) entails

$$\begin{aligned}
&\left\| |\Pi_\tau u|^{4\sigma-1} |\nabla \Pi_\tau u|^2 \right\|_{L^{q_0'}(\mathbb{R}_+;L^{r_0'})} \\
&\lesssim \tau^{-1/2} \|\Pi_\tau u\|^{4\sigma-1}_{L^{q_2}(\mathbb{R}_+;W^{1,r_2})} \|\nabla \Pi_\tau u\|^2_{L^{q_2}(\mathbb{R}_+;W^{1,r_2})} \lesssim \tau^{-1/2} \|u\|^{4\sigma+1}_{Z(\mathbb{R}_+)}.
\end{aligned}$$

Substituting this into (4.2), we arrive at

$$\|\Delta \mathcal{B}_1(u)\|_{L^{q_0'}(\mathbb{R}_+;L^{r_0'})} \lesssim \|u\|^{2\sigma+1}_{Z(\mathbb{R}_+)} + \|u\|^{4\sigma+1}_{Z(\mathbb{R}_+)}. \tag{4.4}$$

For the second term on the right-hand side of (4.1), similar to (3.10), we have

$$\begin{aligned}
\|\partial_t \mathcal{B}_1(u)\|_{L^{q_0'}(\mathbb{R}_+;L^{r_0'})} &= \left\| \partial_t \left(\frac{N(\tau) - \mathbf{1}}{\tau} \Pi_\tau u \right) \right\|_{L^{q_0'}(\mathbb{R}_+;L^{r_0'})} \\
&\lesssim \left\| |\Pi_\tau u|^{2\sigma} \partial_t \Pi_\tau u \right\|_{L^{q_0'}(\mathbb{R}_+;L^{r_0'})}.
\end{aligned}$$

Observing that ∂_t and Π_τ commute, we find from (1.1) that

$$\begin{aligned}
\|\partial_t \mathcal{B}_1(u)\|_{L^{q_0'}(\mathbb{R}_+;L^{r_0'})} &\lesssim \left\| |\Pi_\tau u|^{2\sigma} \Pi_\tau \Delta u \right\|_{L^{q_0'}(\mathbb{R}_+;L^{r_0'})} \\
&\quad + \left\| |\Pi_\tau u|^{2\sigma} \Pi_\tau \left(|u|^{2\sigma} u \right) \right\|_{L^{q_0'}(\mathbb{R}_+;L^{r_0'})}.
\end{aligned}$$

For the first term on the right side, Hölder's inequality (2.4) and Lemma 10 imply

$$\begin{aligned}\left\||\Pi_\tau u|^{2\sigma}\Pi_\tau \Delta u\right\|_{L^{q_0'}(\mathbb{R}_+;L^{r_0'})} &\leq \|\Pi_\tau u\|^{2\sigma}_{L^\gamma(\mathbb{R}_+;L^{r_0})}\|\Pi_\tau \Delta u\|_{L^{q_0}(\mathbb{R}_+;L^{r_0})}\\ &\lesssim \|u\|^{2\sigma}_{L^\gamma(\mathbb{R}_+;L^{r_0})}\|\Delta u\|_{L^{q_0}(\mathbb{R}_+;L^{r_0})}\\ &\lesssim \|u\|^{2\sigma+1}_{Z(\mathbb{R}_+)}. \qquad (4.5)\end{aligned}$$

For the second term, applying Hölder's inequality and (3.8) yields

$$\begin{aligned}&\left\||\Pi_\tau u|^{2\sigma}\Pi_\tau\left(|u|^{2\sigma}u\right)\right\|_{L^{q_0'}(\mathbb{R}_+;L^{r_0'})}\\ &\leq \|\Pi_\tau u\|^{2\sigma}_{L^{q_2}(\mathbb{R}_+;L^{(4\sigma+1)r_0'})}\left\||\Pi_\tau\left(|u|^{2\sigma}u\right)\right\|_{L^{\frac{q_2}{2\sigma+1}}(\mathbb{R}_+;L^{\frac{4\sigma+1}{2\sigma+1}r_0'})}\\ &\leq \|u\|^{4\sigma+1}_{L^{q_2}(\mathbb{R}_+;L^{(4\sigma+1)r_0'})},\end{aligned}$$

where we recall that $q_2 = (4\sigma+1)q_0'$. We verify that under our assumptions on σ,

$$W^{2,r_2}(\mathbb{R}^d) \hookrightarrow L^{(4\sigma+1)r_0'}(\mathbb{R}^d),$$

where (q_2, r_2) is admissible, since $W^{s,r_2}(\mathbb{R}^d) \hookrightarrow L^{(4\sigma+1)r_0'}(\mathbb{R}^d)$ holds for

$$s = d\left(\frac{1}{r_2} - \frac{1}{(4\sigma+1)r_0'}\right) = \frac{d}{2} - \frac{d}{4\sigma+1}\left(\frac{2}{d}+\frac{1}{2}\right) \in [0,2].$$

Consequently, we obtain

$$\left\||\Pi_\tau u|^{2\sigma}\Pi_\tau\left(|u|^{2\sigma}u\right)\right\|_{L^{q_0'}(\mathbb{R}_+;L^{r_0'})} \lesssim \|u\|^{4\sigma+1}_{L^{q_2}(\mathbb{R}_+;W^{2,r_2})} \lesssim \|u\|^{4\sigma+1}_{Z(\mathbb{R}_+)},$$

which, when combined with (4.5), yields

$$\|\partial_t \mathcal{B}_1(u)\|_{L^{q_0'}(\mathbb{R}_+;L^{r_0'})} \lesssim \|u\|^{2\sigma+1}_{Z(\mathbb{R}_+)} + \|u\|^{4\sigma+1}_{Z(\mathbb{R}_+)}.$$

Together with (4.1) and (4.4), this leads to

$$\begin{aligned}\|\mathcal{A}_1(u)\|_{\ell^q(\tau\mathbb{N};L^r)} &\lesssim \tau\|\Delta\mathcal{B}_1(u)\|_{L^{q_0'}(\mathbb{R}_+;L^{r_0'})} + \tau\|\partial_t\mathcal{B}_1(u)\|_{L^{q_0'}(\mathbb{R}_+;L^{r_0'})}\\ &\lesssim \tau\left(\|u\|^{2\sigma+1}_{Z(\mathbb{R}_+)} + \|u\|^{4\sigma+1}_{Z(\mathbb{R}_+)}\right). \qquad (4.6)\end{aligned}$$

We now proceed to estimate $\mathcal{A}_2(u)$. We perform the decomposition:

$$\mathcal{B}_1(u) + i|u|^{2\sigma}u = \mathcal{B}_2(u) + i\mathcal{B}_3(u),$$

where

$$\mathcal{B}_2(u) := \mathcal{B}_1(u) + i|\Pi_\tau u|^{2\sigma}\Pi_\tau u, \quad \mathcal{B}_3(u) := |u|^{2\sigma}u - |\Pi_\tau u|^{2\sigma}\Pi_\tau u.$$

The inequality (3.4) tells us, for admissible pair (q, r),

$$\begin{aligned}&\|\mathcal{A}_2(u)\|_{\ell^q(\tau\mathbb{N};L^r)}\\ &\leq \left\|\int_0^{n\tau} S_\tau(n\tau-s)\mathcal{B}_2(u)ds\right\|_{\ell^q(\tau\mathbb{N};L^r)} + \left\|\int_0^{n\tau} S_\tau(n\tau-s)\mathcal{B}_3(u)ds\right\|_{\ell^q(\tau\mathbb{N};L^r)}\\ &\lesssim \|\mathcal{B}_2(u)\|_{L^{q_0'}(\mathbb{R}_+;L^{r_0'})} + \|\mathcal{B}_3(u)\|_{L^{q_0'}(\mathbb{R}_+;L^{r_0'})}.\end{aligned}$$

Recall that $N(\tau)z = ze^{-i\tau|z|^{2\sigma}}$; thus, the Taylor formula gives the pointwise estimate

$$|\mathcal{B}_2(u)| = \left|\frac{N(\tau)-\mathbf{1}}{\tau}\Pi_\tau u + i|\Pi_\tau u|^{2\sigma}\Pi_\tau u\right| \lesssim \tau|\Pi_\tau u|^{4\sigma+1}.$$

By employing a similar approach as above, we obtain

$$\begin{aligned}\|\mathcal{B}_2(u)\|_{L^{q_0'}(\mathbb{R}_+;L^{r_0'})} &\lesssim \tau\left\||\Pi_\tau u|^{4\sigma+1}\right\|_{L^{q_0'}(\mathbb{R}_+;L^{r_0'})}\\ &\lesssim \tau\|u\|^{4\sigma+1}_{L^{q_2}(\mathbb{R}_+;L^{(4\sigma+1)r_0'})} \lesssim \tau\|u\|^{4\sigma+1}_{L^{q_2}(\mathbb{R}_+;W^{2,r_2})}\\ &\lesssim \tau\|u\|^{4\sigma+1}_{Z(\mathbb{R}_+)}.\end{aligned}$$

Finally, applying Hölder's inequality (2.4) yields

$$\begin{aligned}\|\mathcal{B}_3(u)\|_{L^{q_0'}(\mathbb{R}_+;L^{r_0'})} &= \left\||\Pi_\tau u|^{2\sigma}\Pi_\tau u - |u|^{2\sigma}u\right\|_{L^{q_0'}(\mathbb{R}_+;L^{r_0'})}\\ \lesssim \Big(\|\Pi_\tau u\|^{2\sigma}_{L^{\gamma}(\mathbb{R}_+;L^{r_0})} &+ \|u\|^{2\sigma}_{L^{\gamma}(\mathbb{R}_+;L^{r_0})}\Big)\|\Pi_\tau u - u\|_{L^{q_0}(\mathbb{R}_+;L^{r_0})}\\ \lesssim \tau\|u\|^{2\sigma}_{Z(\mathbb{R}_+)}\|u\|_{L^{q_0}(\mathbb{R}_+;W^{2,r_0})} &\lesssim \tau\|u\|^{2\sigma+1}_{Z(\mathbb{R}_+)},\end{aligned}$$

where we have used (3.5). Hence,

$$\|\mathcal{A}_2(u)\|_{\ell^q(\tau\mathbb{N};L^r)} \lesssim \tau\left(\|u\|^{2\sigma+1}_{Z(\mathbb{R}_+)} + \|u\|^{4\sigma+1}_{Z(\mathbb{R}_+)}\right),$$

which combines with (4.6) concludes the proof of Lemma 13. □

Proof of Theorem 3. In light of Theorem 6 and the assumptions of Theorem 3, we observe that u and Z_τ satisfy the following estimates:

$$\begin{aligned}&\|u\|_{Z(\mathbb{R}_+)} \le M_1,\\ &\max_{A\in\{1,\nabla,J\}}\|A(n\tau)Z_\tau(n\tau)\|_{\ell^\infty(\tau\mathbb{N};L^2)} + \|Z_\tau(n\tau)\|_{\ell^{q_0}(\tau\mathbb{N};W^{1,r_0})} \le M.\end{aligned} \tag{4.7}$$

Instead of estimating $Z_\tau(n\tau)-u(n\tau)$, we shall estimate $Z_\tau(n\tau)-\Pi_\tau u(n\tau)$, as indicated by (3.5):

$$\|u(n\tau)-\Pi_\tau u(n\tau)\|_{\ell^\infty(\tau\mathbb{N};L^2)} \le C\tau\|u(n\tau)\|_{\ell^\infty(\tau\mathbb{N};H^2)} \le C\tau M_1.$$

We recall that $\gamma<\infty$, as defined by (2.3), which satisfies $\gamma\delta(r_0)>1$ (see (2.12)). Therefore, in view of (1.5), (4.7) and (2.6), for any $\varepsilon>0$, we can find a *finite* number $K=K(\varepsilon)$ of time intervals $I_j=\overline{[m_j,m_{j+1})}$ with integers $0=m_1<m_2<\ldots<m_K<m_{K+1}=\infty$, and $\tau(\varepsilon)>0$ such that for $1\le j\le K$ and $0<\tau\le\tau(\varepsilon)$, the following holds

$$\|u(k\tau)\|_{\ell^\gamma(\tau I_j;L^{r_0})} + \|Z_\tau(k\tau)\|_{\ell^\gamma(\tau I_j;L^{r_0})} \le \varepsilon, \quad \mathbb{R}_+ = \bigcup_{j=1}^K I_j. \tag{4.8}$$

On each interval I_j, by combining the discrete Duhamel's formula (3.16) with (2.1), we obtain the following decomposition:

$$Z_\tau(m_j\tau + n\tau) - \Pi_\tau u(m_j\tau + n\tau) = \mathcal{A}_1(n) + \mathcal{A}_2(n) + \mathcal{A}_3(n) + \mathcal{A}_4(n), \quad (4.9)$$

where

$$\begin{aligned}
\mathcal{A}_1(n) &:= S_\tau(n\tau)\left(Z_\tau(m_j\tau) - \Pi_\tau u(m_j\tau)\right), \\
\mathcal{A}_2(n) &:= S_\tau(n\tau)\left(\Pi_\tau u(m_j\tau) - u(m_j\tau)\right), \\
\mathcal{A}_3(n) &:= \tau \sum_{k=0}^{n-1} S_\tau(n\tau - k\tau)\Big(\frac{N(\tau) - \mathbf{1}}{\tau} Z_\tau(m_j\tau + k\tau) \\
&\quad - \frac{N(\tau) - \mathbf{1}}{\tau} \Pi_\tau u(m_j\tau + k\tau)\Big), \\
\mathcal{A}_4(n) &:= \tau \sum_{k=0}^{n-1} S_\tau(n\tau - k\tau) \frac{N(\tau) - \mathbf{1}}{\tau} \Pi_\tau u(m_j\tau + k\tau) \\
&\quad + i \int_0^{n\tau} S_\tau(n\tau - s)\left(|u|^{2\sigma} u\right)(m_j\tau + s) ds,
\end{aligned}$$

and we omit the dependence of the $\mathcal{A}_k$ terms on j for the sake of brevity. The terms $\mathcal{A}_1$ and $\mathcal{A}_2$ are linear, while $\mathcal{A}_3$ and $\mathcal{A}_4$ are nonlinear. The goal is to demonstrate that the term $\mathcal{A}_3$ can be absorbed by the left-hand side of (4.9), $\mathcal{A}_2$ and $\mathcal{A}_4$ are $\mathcal{O}(\tau)$, and $\mathcal{A}_1$ can be estimated by induction on j.

Let $(q, r) \in \{(q_0, r_0), (\infty, 2)\}$. The homogeneous Strichartz estimate (3.2) yields

$$\|\mathcal{A}_1\|_{\ell^q(\tau I_j; L^r)} \le C_{d,q} \|Z_\tau(m_j\tau) - \Pi_\tau u(m_j\tau)\|_{L^2}.$$

The second term is controlled via (3.2) and (3.5), leading to

$$\begin{aligned}
\|\mathcal{A}_2\|_{\ell^q(\tau I_j; L^r)} &= \|S_\tau(n\tau)\left(\Pi_\tau u(m_j\tau) - u(m_j\tau)\right)\|_{\ell^q(\tau I_j; L^r)} \\
&\le C_{d,q} \|\Pi_\tau u(m_j\tau) - u(m_j\tau)\|_{L^2} \\
&\le C_{d,q}\tau \|u(m_j\tau)\|_{H^2} \le C_{d,q}\tau M_1,
\end{aligned}$$

where we have invoked (4.7). To estimate $\mathcal{A}_3$, we apply (3.12) to obtain

$$\left|\frac{N(\tau) - \mathbf{1}}{\tau} Z_\tau - \frac{N(\tau) - \mathbf{1}}{\tau} \Pi_\tau u\right| \lesssim \left(|Z_\tau|^{2\sigma} + |\Pi_\tau u|^{2\sigma}\right)|Z_\tau - \Pi_\tau u|.$$

The inhomogeneous Strichartz estimate (3.3) in conjunction with Hölder's inequality (2.4) provides

$$\begin{aligned}
\|\mathcal{A}_3\|_{\ell^q(\tau I_j; L^r)} &\lesssim \left\|\left(|Z_\tau|^{2\sigma} + |\Pi_\tau u|^{2\sigma}\right)|Z_\tau - \Pi_\tau u|\right\|_{\ell^{q_0'}(\tau I_j; L^{r_0'})} \\
&\lesssim \left(\|Z_\tau\|^{2\sigma}_{\ell^\gamma(\tau I_j; L^{r_0})} + \|\Pi_\tau u\|^{2\sigma}_{\ell^\gamma(\tau I_j; L^{r_0})}\right)\|Z_\tau - \Pi_\tau u\|_{\ell^{q_0}(\tau I_j; L^{r_0})} \\
&\le C_{d,\sigma,q}\varepsilon^{2\sigma}\|Z_\tau - \Pi_\tau u\|_{\ell^{q_0}(\tau I_j; L^{r_0})},
\end{aligned}$$

for some constant $C_{d,\sigma,q}$, where we have used (3.6) and (4.8).

We now choose $\varepsilon > 0$ sufficiently small such that $C_{d,\sigma,q_0}\varepsilon^{2\sigma} < 1/2$; thus, we get

$$\|\mathcal{A}_3\|_{\ell^{q_0}(\tau I_j;L^{r_0})} \leq \frac{1}{2}\|Z_\tau - \Pi_\tau u\|_{\ell^{q_0}(\tau I_j;L^{r_0})}, \quad 1 \leq j \leq K.$$

Lemma 13 indicates that

$$\max_{1\leq j\leq K} \left(\|\mathcal{A}_4\|_{\ell^\infty(\tau I_j;L^2)} + \|\mathcal{A}_4\|_{\ell^{q_0}(\tau I_j;L^{r_0})}\right) \lesssim \tau. \tag{4.10}$$

Collecting all previous estimates, we first get, with $(q,r) = (q_0, r_0)$,

$$\begin{aligned}\|Z_\tau - \Pi_\tau u\|_{\ell^{q_0}(\tau I_j;L^{r_0})} \leq C \left\|Z_\tau(m_j\tau) - \Pi_\tau u(m_j\tau)\right\|_{L^2} \\ + C\tau + \frac{1}{2}\|Z_\tau - \Pi_\tau u\|_{\ell^{q_0}(\tau I_j;L^{r_0})},\end{aligned}$$

leading to

$$\|Z_\tau - \Pi_\tau u\|_{\ell^{q_0}(\tau I_j;L^{r_0})} \leq 2C \left\|Z_\tau(m_j\tau) - \Pi_\tau u(m_j\tau)\right\|_{L^2} + 2C\tau, \quad 1 \leq j \leq K.$$

Next taking $(q,r) = (\infty, 2)$, we find

$$\begin{aligned}&\|Z_\tau - \Pi_\tau u\|_{\ell^\infty(\tau I_j;L^2)} \\ &\lesssim \left\|Z_\tau(m_j\tau) - \Pi_\tau u(m_j\tau)\right\|_{L^2} + \tau + \|Z_\tau - \Pi_\tau u\|_{\ell^{q_0}(\tau I_j;L^{r_0})} \\ &\lesssim \left\|Z_\tau(m_j\tau) - \Pi_\tau u(m_j\tau)\right\|_{L^2} + \tau, \quad 1 \leq j \leq K.\end{aligned}$$

By construction, $m_1 = 0$, so $Z_\tau(m_1\tau) - \Pi_\tau u(m_1\tau) = 0$. For $2 \leq j \leq K$, we have

$$\left\|Z_\tau(m_j\tau) - \Pi_\tau u(m_j\tau)\right\|_{L^2} \leq \|Z_\tau - \Pi_\tau u\|_{\ell^\infty(\tau I_{j-1};L^2)},$$

thus, we derive the first conclusion in Theorem 3 by the finiteness of K. The second conclusion follows directly from the first conclusion and the scattering result recalled in Theorem 6: there exists $u_+ \in \Sigma$ such that

$$\|u(t) - S(t)u_+\|_{L^2} \underset{t\to\infty}{\longrightarrow} 0,$$

where we used the fact that $S(t)$ is unitary on L^2. □

A.1. Continuous and discrete Strichartz estimates

First, we recall the main steps of the proof of Proposition 4, as presented in [20]. By duality, the homogeneous estimate is equivalent to:

$$\left\|\int S(-t)F(t)dt\right\|_{L^2(\mathbb{R}^d)} \lesssim \|F\|_{L^{q'}(\mathbb{R};L^{r'}(\mathbb{R}^d))}. \tag{A.1}$$

By the TT^* method, this is equivalent to

$$\left|\iint \langle S(-t_2)F(t_2), S(-t_1)G(t_1)\rangle \, dt_2 dt_1\right| \lesssim \|F\|_{L^{q'}L^{r'}} \|G\|_{L^{q'}L^{r'}} .$$

By symmetry, it suffices to prove

$$|T(F,G)| \lesssim \|F\|_{L^{q'}L^{r'}} \|G\|_{L^{q'}L^{r'}} , \tag{A.2}$$

where $T(F,G) = \iint_{t_2<t_1} \langle S(-t_2)F(t_2), S(-t_1)G(t_1)\rangle \, dt_2 dt_1$. Since S is unitary on L^2,

$$|\langle S(-t_2)F(t_2), S(-t_1)G(t_1)\rangle| \leq \|F(t_2)\|_{L^2} \|G(t_1)\|_{L^2} .$$

We infer from the dispersive estimate (see e.g. [1]),

$$\|S(t)\|_{L^1(\mathbb{R}^d)\to L^\infty(\mathbb{R}^d)} \leq \frac{1}{(2\pi|t|)^{d/2}}, \tag{A.3}$$

that

$$\begin{aligned}|\langle S(-t_2)F(t_2), S(-t_1)G(t_1)\rangle| &= |\langle S(t_1-t_2)F(t_2), G(t_1)\rangle| \\ &\lesssim \frac{1}{|t_2-t_1|^{d/2}} \|F(t_2)\|_{L^1} \|G(t_1)\|_{L^1} .\end{aligned}$$

By interpolation,

$$|\langle S(-t_2)F(t_2), U(-t_1)G(t_1)\rangle| \lesssim \frac{1}{|t_2-t_1|^{2/q}} \|F(t_2)\|_{L^{r'}} \|G(t_1)\|_{L^{r'}} ,$$

since (q,r) is admissible. In view of future comparison with the discrete case, we recall the Hardy–Littlewood–Sobolev inequality, which can be found e.g. in [21, Sect. 4.3] or [22, Th. 0.3.2], in the one-dimensional case corresponding to our time variable:

Proposition 14 (Hardy-Littlewood-Sobolev inequality). *For $\alpha \in (0,1)$, define the fractional integral operator by*

$$I_\alpha f(t) = \int_{\mathbb{R}} |t-s|^{-\alpha} f(s) ds.$$

If, for some $1 < p < q < \infty$,

$$\alpha + \frac{1}{p} = 1 + \frac{1}{q},$$

then there exists $C_{p,q}$ such that

$$\|I_\alpha f\|_{L^q(\mathbb{R})} \leq C_{p,q} \|f\|_{L^p(\mathbb{R})}, \quad \forall f \in L^p(\mathbb{R}).$$

The Hardy–Littlewood–Sobolev inequality can be viewed as a generalized Young inequality, owing to the fact that the map $t \mapsto \frac{1}{t}$ belongs to the weak L^1 space $L^1_w(\mathbb{R})$ (see [23] for generalizations). Then (A.2) follows from the above inequality, with $\alpha = 2/q$ and $p = q'$. This proves the homogeneous estimate.

For the inhomogeneous case, consider two admissible pairs (q,r) and $(\widetilde{q},\widetilde{r})$. By duality, (2.2) is equivalent to:

$$|T(F,G)| \lesssim \|F\|_{L^{q'}_t L^{r'}_x} \|G\|_{L^{\widetilde{q}'}_t L^{\widetilde{r}'}_x}. \tag{A.4}$$

We have

$$|T(F,G)| \leq \left(\sup_{t\in\mathbb{R}} \left\| \int_{t_2<t_1} S(-t_2)F(t_2)dt_2 \right\|_{L^2_x} \right) \|G\|_{L^1_t L^2_x},$$

and when $(\widetilde{q},\widetilde{r}) = (\infty, 2)$, (A.4) follows from (A.1). Similarly, (A.4) holds when $(q,r) = (\infty,2)$. From (A.2), one has (A.4) when $(q,r) = (\widetilde{q},\widetilde{r})$. The general case (2.2) follows by interpolation between these three cases.

If the time variable becomes discrete, the above approach fails. First, as noticed in [7, 10], the homogeneous estimate from Proposition 4 cannot be true when the L^q-norm is replaced by the discrete ℓ^q-norm as defined in (1.5). Indeed, for finite q, this would imply

$$\tau^{1/q}\|S(\tau)\phi\|_{L^r(\mathbb{R}^d)} \lesssim \|\phi\|_{L^2(\mathbb{R}^d)}, \quad \forall \phi \in L^2(\mathbb{R}^d).$$

Choosing $\phi = U(-\tau)f$ with $f \in L^2(\mathbb{R}^d) \setminus L^r(\mathbb{R}^d)$ rules out such an inequality. Therefore, as noticed in [11], a frequency cutoff must be introduced, involving a CFL condition. In [7], like in the present contribution, the choice $|\xi| \lesssim \tau^{-1/2}$ is made. Other choices may lead to different estimates, as shown in [12].

With $S_\tau(t) = S(t)\Pi_\tau = \Pi_\tau S(t)$, where Π_τ is the Fourier multiplier defined by (1.4) and $\chi \in C_c^\infty(\mathbb{R}^d)$, the discrete Strichartz estimates stated in Proposition 7 become true. Note that if S_τ satisfied a dispersive inequality of the form (A.3), we would face a serious technical issue when invoking the discrete Hardy-Littlewood-Sobolev inequality, as in the discrete case, the singularity at the origin becomes fatal. Recall the result from [24]:

Lemma 15. *Let $0 < \alpha < 1$, and consider a kernel satisfying*

$$|k(n)| \leq \frac{C}{1+|n|^\alpha}, \quad n \in \mathbb{Z}. \tag{A.5}$$

Then the operator I defined by $If = f * k$ is continuous from $\ell^p(\mathbb{Z})$ to $\ell^q(\mathbb{Z})$ provided that $1 < p < q < \infty$ and

$$\alpha + \frac{1}{p} = 1 + \frac{1}{q},$$

that is,

$$\|If\|_{\ell^q(\mathbb{Z})} \leq C\|f\|_{\ell^p(\mathbb{Z})}, \quad \forall f \in \ell^p(\mathbb{Z}).$$

The introduction of the frequency cutoff Π_τ indeed turns the estimate (A.3) into an estimate of the form (A.5), since for all $\phi \in \mathcal{S}(\mathbb{R}^d)$,

$$\|S_\tau(t)\phi\|_{L^\infty(\mathbb{R}^d)} \leq \frac{C_d}{\tau^{d/2} + |t|^{d/2}} \|\phi\|_{L^1(\mathbb{R}^d)}, \quad \forall t \in \mathbb{R}.$$

This estimate follows directly from Young inequality, since Π_τ may be viewed as a convolution operator with a smooth kernel (depending on τ). See e.g. [12]. Since in addition $\|S_\tau(t)\phi\|_{L^2} \leq \|\phi\|_{L^2}$, the proof of Proposition 7 is essentially the same as above.

References

[1] T. Cazenave, *Semilinear Schrödinger equations.* vol. 10, *Courant Lecture Notes in Mathematics*, New York University Courant Institute of Mathematical Sciences, New York (2003).

[2] J. Ginibre and G. Velo, On a class of nonlinear Schrödinger equations. I The Cauchy problem, general case, *J. Funct. Anal.* **32**, 1–32 (1979).

[3] J. Ginibre and G. Velo, On a class of nonlinear Schrödinger equations. II Scattering theory, general case, *J. Funct. Anal.* **32**, 33–71 (1979).

[4] C. Besse, B. Bidégaray, and S. Descombes, Order estimates in time of splitting methods for the nonlinear Schrödinger equation, *SIAM J. Numer. Anal.* **40**(1), 26–40 (2002).

[5] C. Lubich, On splitting methods for Schrödinger-Poisson and cubic nonlinear Schrödinger equations, *Math. Comp.* **77**(264), 2141–2153 (2008).

[6] J. Ginibre and G. Velo, The global Cauchy problem for the nonlinear Schrödinger equation revisited, *Ann. Inst. H. Poincaré Anal. Non Linéaire.* **2**, 309–327 (1985).

[7] L. I. Ignat, A splitting method for the nonlinear Schrödinger equation, *J. Differ. Equations.* **250**(7), 3022–3046 (2011).

[8] L. I. Ignat and E. Zuazua. Dispersive properties of numerical schemes for nonlinear Schrödinger equations. In *Foundations of computational mathematics, Santander 2005. Selected papers based on the presentations at the international conference of the Foundations of Computational Mathematics (FoCM), Santander, Spain, June 30 – July 9, 2005.*, pp. 181–207. Cambridge: Cambridge University Press (2006).

[9] L. I. Ignat and E. Zuazua, Numerical dispersive schemes for the nonlinear Schrödinger equation, *SIAM J. Numer. Anal.* **47**(2), 1366–1390 (2009).

[10] A. Stefanov and P. G. Kevrekidis, Asymptotic behaviour of small solutions for the discrete nonlinear Schrödinger and Klein-Gordon equations, *Nonlinearity.* **18**(4), 1841–1857 (2005).

[11] E. Faou and B. Grébert, Hamiltonian interpolation of splitting approximations for nonlinear PDEs, *Found. Comput. Math.* **11**(4), 381–415 (2011).

[12] A. Ostermann, F. Rousset, and K. Schratz, Error estimates of a Fourier integrator for the cubic Schrödinger equation at low regularity, *Found. Comput. Math.* **21**(3), 725–765 (2021).
[13] W. Choi and Y. Koh, On the splitting method for the nonlinear Schrödinger equation with initial data in H^1, *Discrete Contin. Dyn. Syst.* **41**(8), 3837–3867 (2021).
[14] E. Faou, R. Horsin, and F. Rousset, On numerical Landau damping for splitting methods applied to the Vlasov-HMF model, *Found. Comput. Math.* **18**(1), 97–134 (2018).
[15] R. Carles and C. Su, Scattering and uniform in time error estimates for splitting method in NLS, *Found. Comput. Math.* **24**(2), 683–722 (2024).
[16] T. Tao, *Nonlinear dispersive equations.* vol. 106, *CBMS Regional Conference Series in Mathematics*, Published for the Conference Board of the Mathematical Sciences, Washington, DC (2006). Local and global analysis.
[17] T. Cazenave and F. Weissler, Rapidly decaying solutions of the nonlinear Schrödinger equation, *Comm. Math. Phys.* **147**, 75–100 (1992).
[18] K. Nakanishi and T. Ozawa, Remarks on scattering for nonlinear Schrödinger equations, *NoDEA Nonlinear Differential Equations Appl.* **9**(1), 45–68 (2002).
[19] N. Hayashi and Y. Tsutsumi. Remarks on the scattering problem for nonlinear Schrödinger equations. In *Differential equations and mathematical physics (Birmingham, Ala., 1986)*, vol. 1285, *Lectures Notes in Math.*, pp. 162–168. Springer, Berlin (1987).
[20] M. Keel and T. Tao, Endpoint Strichartz estimates, *Amer. J. Math.* **120**(5), 955–980 (1998).
[21] E. H. Lieb and M. Loss, *Analysis*, second edn. vol. 14, *Graduate Studies in Mathematics*, American Mathematical Society, Providence, RI (2001).
[22] C. D. Sogge, *Fourier integrals in classical analysis*, second edn. vol. 210, *Cambridge Tracts in Mathematics*, Cambridge University Press, Cambridge (2017).
[23] E. M. Stein, *Singular integrals and differentiability properties of functions.* Princeton Mathematical Series, No. 30, Princeton University Press, Princeton, N.J. (1970).
[24] E. M. Stein and S. Wainger, Discrete analogues in harmonic analysis. II. Fractional integration, *J. Anal. Math.* **80**, 335–355 (2000).

https://doi.org/10.1142/9789819816620_0005

Chapter 5

Modulated Solitons Algorithms for Nonlinear Schrödinger Equations

Erwan Faou*, Yoann Le Hénaff† and Pierre Raphaël‡

*Univ Rennes, INRIA Bretagne, IRMAR-UMR 6625, F-35000 Rennes, France
erwan.faou@inria.fr

†Mathematisches Institut Universität Tübingen Auf der Morgenstelle 10 D-72076 Tübingen, Germany
yoann.le-henaff@uni-tuebingen.de

‡Department of Pure Mathematics and Mathematical Statistics, University of Cambridge, Cambridge, U.K.
pr463@dpmms.cam.ac.uk

We discuss in this chapter new algorithms for the nonlinear cubic Schrödinger equation, based on the representation by modulated solitons in interactions. We consider the general idea of modulation and calculate explicitly the evolution of a single soliton for which the evolution system of the modulation parameters forms a completely integrable system. We then discuss the numerical approximation of nonlinear interactions of several modulated solitons through the Dirac-Frenkel variational principle. Numerical examples illustrate that the method yields very satisfying results in the case of weak interactions, and allows the computations of interacting solitons in the whole space without a full grid of discretization.

Contents

1. Introduction

We consider in this work the cubic nonlinear Schrödinger equation with harmonic potential:

$$\begin{cases} i\partial_t u(t,x) = -\Delta u(t,x) + |x|^2 u(t,x) + \mu |u(t,x)|^2 u(t,x), \\ u(0,\cdot) = U_0, \end{cases} \tag{1.1}$$

where $(t,x) \in \mathbb{R}_+ \times \mathbb{R}^d$, $\Delta = \frac{\partial^2}{\partial x_1^2} + \cdots + \frac{\partial^2}{\partial x_d^2}$, $|x|^2 = x_1^2 + \cdots + x_d^2$ and $\mu \in \mathbb{R}$, and U_0 a smooth given function. In the case $\mu < 0$ the equation is said to be *focusing*, and if $\mu > 0$ it is *defocusing*. When $\mu = 0$ it is the well-known harmonic oscillator. The cubic nonlinearity is treated for its physical importance, but the method presented should be adaptable to any power-type nonlinearity. Equation (1.1) is also sometimes called the "Gross-Pitaevskii" equation.

The relevance of the Schrödinger equation in a multi-dimensional context arises from its physical nature. For instance, if one considers the variable space of two coherent particles moving in three-dimensional space, we get $d = 6$ (see [1]). We also refer to [2] for a good introduction to the high-dimensional linear Schrödinger equation in molecular dynamics (i.e. in the case $\mu = 0$ but with potentials with complex structures). The cubic nonlinearity is usually used when studying the Bose-Einstein condensates in the two- or three-dimensional context. The nonlinear Schrödinger equation can also be recovered from the Maxwell equation in specific settings, for instance in the field propagation of a single-mode optical fiber [3, 4]. We refer to [5, 6] for extensive studies of the nonlinear Schrödinger equation with general local nonlinearities.

There has been also many works dedicated to the numerical approximation of (1.1). Grid-based methods are very efficient at recovering the effects of the nonlinearity because they do not assume any information about the solution, but they are also very difficult to use in practice since the numerical grids need to be very fine, see e.g. [7–10]. Moreover, a space discretization has to be combined with a time discretization and time splitting method are usually employed, see [11–13]. In molecular dynamics, grid-free base methods have been extensively studied by considering approximation of (1.1) by modulated Gaussian. This is very efficient and widely applied in the case of the linear Schrödinger equation with potential, in particular in problems coming from molecular dynamics (see e.g. [2, 14–18]).

In the nonlinear situation, the computation of *interacting solitons* has known many interesting progresses in physics and in numerical analysis [9, 19]. Similarly, from the theoretical point of view very recent works [20, 21] suggest to consider solutions ψ of (1.1) that are sum of N modulated functions, which write as:

$$\psi(t,x) \approx u(t,x) := \sum_{j=1}^{N} u^j(t,x), \tag{1.2}$$

where

$$u^j(t,x) := \frac{1}{L^j} e^{i\gamma^j + iL^j\beta^j \cdot y^j - i\frac{B^j}{4}|y^j|^2} v^j(t,y^j), \quad y^j := \frac{x - X^j}{L^j}, \tag{1.3}$$

and $N \in \mathbb{N}^*$. Such a representation means that we seek the solution of the equation on a manifold made of modulated functions v^j. The time dependence of the parameters $L^j, B^j, X^j, \beta^j, \gamma^j$ has not been written in (1.3) for the sake of clarity, but it is one of the main ingredients of the approach. More precisely, the core idea is to plug the ansatz (1.2) into (1.1) in order to obtain ODEs for the parameters by projection onto the manifold.

The choice of the function v^j is fundamental. In the linear case, *i.e.* when the cubic nonlinearity is replaced by some multiplication with a potential, and particularly in the semi-classical regime, the v^j are chosen as Gaussian functions. Such techniques have been called *Variational Gaussian wave packets* and have been extensively analyzed, see for instance [2], with many refinements and application in the field of Chemical Physics [14, 15, 22–24].

In the nonlinear setting, we can also choose the v^j Gaussian as in the linear case, and consider the approximation by the Dirac-Frenkel principle, but a more natural choice consists in taking v^j as solitons *i.e.* functions Q solution of the following eigenproblem:

$$-\Delta Q(x) + |x|^2 Q(x) + \mu |Q(x)|^2 Q(x) = \omega Q(x), \quad x \in \mathbb{R}^d, \tag{1.4}$$

for which some symmetry can be imposed (typically radially symmetric). Note that ω is essentially piloted by the L^2 norm of Q, that can be chosen arbitrarily. We refer to [25, 26] for the existence theory of such solution. In particular we can find radial solutions that are positive (ground states) by variational or ordinary differential equation techniques. We can also construct for $|\mu| \ll 1$ solutions to (1.4) that are close to Gaussian or Hermite functions when $\mu \to 0$ by bifurcation and the Lyapunov-Schmidt method. Considering a manifold of the form (1.2) made of modulated solitons $v^j := Q^j$ for some j is thus a natural generalization to the nonlinear case of the classical Gaussian algorithms used in the linear case and in molecular dynamics.

Let us now explain the main ideas underlying the full modulation (1.3) – developed in various works, see for instance [20, 21] and the references therein – and why it is particularly adapted to the nonlinear case.

Consider for instance the case of one function v, *i.e.* $N = 1$. When plugging the ansatz (1.3) into (1.1), we obtain an equation of the form

$$L^2 i\partial_t v + \Delta_y v - |y|^2 v - \mu|v|^2 v + R(t; y, \partial_y)v = 0, \tag{1.5}$$

where $R(t; y, \partial_y)$ is a quadratic operator in y and ∂_y, which depends on time t through the parameters $P = (\gamma, X, \beta, B, L)$ and their time derivatives with respect to t. See (2.4) for more precise detail. It is then possible to choose the parameters in such a way that for instance $R(t; y, \partial_y)v = \omega v$ for some $\omega \in \mathbb{R}$, and to take $v = Q$ a soliton solution of the stationary equation (1.4) which will be a constant in time solution of (1.5). This yields a differential system to be solved by the parameters $P = (\gamma, X, \beta, B, L)$ which is given below by (2.10). It turns out that these equations form a *completely integrable Poisson system* that can be solved explicitly.

This kind of approach has been used successfully in various situations from a theoretical point of view, see [20, 21, 27, 28] and the references therein. Typically, when $N \geq 2$, several modulated solitons interact and this can produce finite time blow-up, periodic or quasi-periodic in time solutions, or growth of Sobolev norm phenomena. A large part of the analysis relies on the ability of calculating nonlinear interactions between two modulated solitons that are possibly weakly interacting. Note that the calculation of soliton interaction can be done explicitly for instance in an integrable situation, e.g. the Szegö equation [29]. But in the general case, we have to rely on numerical simulations.

Another by-product of these modulation techniques in 2D is to make a link between (1.1) on a finite time interval and the Schrödinger equation without harmonic potential

$$i\partial_s \psi + \Delta_x \psi = \mu \psi |\psi|^2, \quad x \in \mathbb{R}^d \tag{1.6}$$

on an unbounded time interval. In this case, the modulation equations generate the so-called *lens* transform, see for instance [6]. Note that our algorithms could also be applied to the latter equation but we will restrict our analysis to the Harmonic case. Let us note as well that such modulation techniques can also be related with the families of exact splitting introduced in [13].

In this work, we retain the idea of approximating solutions to (1.1) by modulating the parameters L^j, B^j, X^j, β^j, γ^j in combination with the Dirac-Frenkel principle consisting in projecting the equation on the manifold (1.3).

As we have explained above, for $N = 1$ the equation will be exact up to the numerical discretization of the equation (1.4). In the general case $N \geq 2$, the only difficulty comes from the calculation of the interactions terms: if $u = f + g$, the nonlinearity $|u|^2 u$ can be split into two terms. The first one is $|f|^2 f + |g|^2 g$, and by the soliton definition it can be taken into account "explicitly". The second term is the interaction term of the form $f^2 \bar{g}$ or $|g|^2 f$, or any combination of three solitons in the case $N \geq 3$. To be evaluated, these terms have to be projected on the modulated soliton manifold. We approximate the interaction terms by using the Dirac-Frenkel principle using grid discretization that will behave favorably for weak interactions (when the solitons are distant for instance).

2. General modulation

2.1. *Notations*

We set

$$(f, g)_{L^2} = \int f\bar{g}, \quad \|f\|_{L^2}^2 = (f, f)_{L^2}, \quad \langle f, g\rangle = \mathrm{Re} \int f\bar{g}, \tag{2.1}$$

the latter bracket being the real scalar product on complex functions of $L^2(\mathbb{R}^d)$ viewed as a real Hilbert space. We can find for instance in [5] or [6] that the following (real) quantities are preserved along solutions of (1.1):

- L^2 norm: $\|\psi\|_{L^2}^2$,
- Energy: $\mathcal{E}(t) = \frac{1}{2}\left(-\Delta\psi + |x|^2\psi, \psi\right)_{L^2} + \frac{\mu}{4}\left(|\psi|^2\psi, \psi\right)_{L^2}$
- Momentum: $m(t) = \left(\mathcal{E}(t) - \|x\psi\|_{L^2}^2\right)^2 + \left(\Im\left[\int x \cdot \nabla\psi\bar{\psi}\right]\right)^2$.

Let $X, \beta \in \mathbb{R}^d$, $\gamma, B \in \mathbb{R}$ and $L > 0$. We set

$$P = (\gamma, X, \beta, B, L) \in \mathbb{R}^{2d+2} \times \mathbb{R}_+, \tag{2.2}$$

the set of modulation parameters, and we define an operator $\mathcal{M}_P : L^2(\mathbb{R}^d) \to L^2(\mathbb{R}^d)$, where for $f \in L^2(\mathbb{R}^d)$,

$$(\mathcal{M}_P f)(t,x) := \frac{1}{L} \exp\left\{ i\gamma + iL\beta \cdot y + \frac{iB}{2}|y|^2 \right\} f(t,y), \quad y := \frac{x - X}{L}. \tag{2.3}$$

When several different modulations are involved, we will denote them by exponents: for instance, $\mathcal{M}_{P^j}$ will be the j-th modulation operator, associated to a set of modulation parameters P^j (see (2.2)). The modulation parameters follow the same rule, and to denote a squared modulation parameter, we will write for example $(L^j)^2$. For a vector, its n-th component will be denoted by the index $_n$. For example X^j_n denotes the n-th component of the modulation parameter X^j. We also write

$$M_P(x) := \frac{1}{L} \exp\left\{ i\gamma + iL\beta \cdot \frac{x - X}{L} + \frac{iB}{2}\left|\frac{x - X}{L}\right|^2 \right\}.$$

Let us remark that we have

$$\|(\mathcal{M}_P f)(x)\|_{L^2}^2 = \frac{1}{L^2} \int_{\mathbb{R}^d} |f|^2(\frac{x}{L}) dx = L^{d-2} \|f\|_{L^2}^2$$

Hence for a given f we can fix the L^2 norm of $(\mathcal{M}_P f)(x)$ by scaling L, except in dimension 2 where this norm has to be fixed by f itself.

2.2. *Single modulation*

We now search u solution of the nonlinear Schrödinger equation as a modulated function v, i.e.

$$u(t,x) = (\mathcal{M}_{P(t)} v)(t,x) = M_{P(t)}(x) v(t, y(t,x)), \quad y(t,x) = \frac{x - X}{L}.$$

We will reformulate the nonlinear Schrödinger equation with unknown $u(t,x)$ as another nonlinear equation with unknown $v(t, y(t,x))$. In this new nonlinear equation the modulation parameters will appear, as well as their time derivative. By correctly choosing the time evolution of these parameters, it will be shown that the function v can simply be expressed.

Proposition 1. *Let $P(t)$ be a set of time dependent parameters (2.2) and $u(t,x) = (\mathcal{M}_{P(t)}v)(t,x)$. We have*

$$
\begin{aligned}
&(i\partial_t u + \Delta_x u - |x|^2 u - \mu|u|^2 u)(t,x) = M_{P(t)}(x)\\
&\times \left[\begin{array}{c} i\partial_t + i\left(-\dfrac{\dot L}{L} + \dfrac{B}{L^2}d\right) - \dot\gamma + \beta\cdot\dot X - |X|^2 - |\beta|^2 \\ + \left(-L\dot\beta + \dfrac{B}{L}\dot X - 2\dfrac{B}{L}\beta - 2LX\right)\cdot y \\ + \left(\dfrac{B\dot L}{L} - \dfrac{\dot B}{2} - \dfrac{B^2}{L^2} - L^2\right)|y|^2 + i\left(2\dfrac{\beta}{L} - \dfrac{\dot X}{L}\right)\cdot\nabla \\ + i\left(2\dfrac{B}{L^2} - \dfrac{\dot L}{L}\right) y\cdot\nabla + \dfrac{1}{L^2}\Delta_y - \dfrac{\mu}{L^2}|v(t,y(t,x))|^2 \end{array}\right] v(t,y(t,x)).
\end{aligned}
\tag{2.4}
$$

Proof. We start by computing, in dimension $d \geq 1$,

$$\Delta_x u = \frac{e^{i\gamma}}{L^2}\Delta_y\left[e^{iL\beta\cdot y + i\frac{B}{2}|y|^2}v(t,y)\right].$$

After differentiating with respect to the n-th coordinate, we get for $n = 1,\ldots,d$,

$$\partial_{y_n}\left[e^{iL\beta\cdot y+i\frac{B}{2}|y|^2}v\right] = e^{iL\beta\cdot y+i\frac{B}{2}|y|^2}\left[\partial_{y_n}v + i\left(L\beta_n + By_n\right)v\right]. \tag{2.5}$$

Differentiating again with respect to the n-th coordinate:

$$
\begin{aligned}
&\partial^2_{y_n}\left[e^{iL\beta\cdot y+i\frac{B}{2}|y|^2}v\right]\\
&= e^{iL\beta\cdot y+i\frac{B}{2}|y|^2}\left[\begin{array}{c}\partial^2_{y_n}v + i\left(L\beta_n + By_n\right)\partial_{y_n}v + iBv \\ + i\left(L\beta_n + By_n\right)\left[\partial_{y_n}v + i\left(L\beta_n + By_n\right)v\right]\end{array}\right]\\
&= e^{iL\beta\cdot y+i\frac{B}{2}|y|^2}\left[\begin{array}{c}\partial^2_{y_n}v + 2i\left(L\beta_n + By_n\right)\partial_{y_n}v \\ + iBv - \left(L\beta_n + By_n\right)^2 v\end{array}\right].
\end{aligned}
\tag{2.6}
$$

By summing (2.6) over $n = 1,\ldots,d$, we get

$$
\begin{aligned}
&\Delta_x u = M_{P(t)}(x)\\
&\times\left[\frac{1}{L^2}\Delta_y v + 2i\left(\frac{\beta}{L} + \frac{B}{L^2}y\right)\cdot\nabla v + i\frac{Bd}{L^2}v - \left|\beta + \frac{B}{L}y\right|^2\right]\\
&= M_{P(t)}(x)\\
&\times\left[\frac{1}{L^2}\Delta_y v + 2i\left(\frac{\beta}{L} + \frac{B}{L^2}y\right)\cdot\nabla v + i\frac{Bd}{L^2}v\right.\\
&\left. - \left(|\beta|^2 + 2\frac{B}{L}\beta\cdot y + \frac{B^2}{L^2}|y|^2\right)v\right].
\end{aligned}
$$

We have

$$-|x|^2 u = -M_{P(t)}(x)\,|Ly+X|^2\,v$$
$$= M_{P(t)}(x)\,(-L^2|y|^2 - 2LX\cdot y - |X|^2)\,v,$$

thus

$$(\Delta_x - |x|^2)u = M_{P(t)}(x) \times \left[\begin{array}{c} \frac{1}{L^2}\Delta_y v + 2i\left(\frac{\beta}{L} + \frac{B}{L^2}y\right)\cdot\nabla v + i\frac{Bd}{L^2}v \\ + \left(-|\beta|^2 - 2\frac{B}{L}\beta\cdot y - \frac{B^2}{L^2}|y|^2 - L^2|y|^2 - 2LX\cdot y - |X|^2\right)v \end{array}\right], \tag{2.7}$$

We now differentiate u with respect to time t:

$$\partial_t u(t,x) = \partial_t\left(e^{i\gamma + iL\beta\cdot y(t,x) + i\frac{B}{2}|y(t,x)|^2}\frac{1}{L}v(t,y(t,x))\right)$$
$$= e^{i\gamma+iL\beta\cdot y+i\frac{B}{2}|y|^2}\left[-\frac{\dot L}{L^2}v(t,y) + \frac{1}{L}\left(\partial_t v + \frac{dy}{dt}\cdot\nabla v\right)\right]$$
$$+ iM_{P(t)}(x)\left[\dot\gamma + L\dot\beta\cdot y + \dot L\beta\cdot y + L\beta\cdot\frac{dy}{dt} + \frac{\dot B}{2}|y|^2 + B\frac{dy}{dt}\cdot y\right]v(t,y).$$

with the notation $y = y(t,x)$ for which we have

$$\frac{dy}{dt} = -\frac{\dot X}{L} - \frac{\dot L}{L}y.$$

Thus

$$\partial_t u(t,x) = M_{P(t)}(x)\left[-\frac{\dot L}{L} + \partial_t - \frac{\dot X}{L}\cdot\nabla - \frac{\dot L}{L}y\cdot\nabla\right]v$$
$$+ iM_{P(t)}(x)\left[\dot\gamma + \left(L\dot\beta - \frac{B\dot X}{L}\right)\cdot y - \beta\cdot\dot X + \left(\frac{\dot B}{2} - \frac{B\dot L}{L}\right)|y|^2\right]v(t,y).$$

Hence,

$$i\partial_t u = M_{P(t)}(x)\left[\begin{array}{c} i\partial_t - i\frac{\dot L}{L} - \dot\gamma + \beta\cdot\dot X \\ + \left(-L\dot\beta + \frac{B}{L}\dot X\right)\cdot y \\ + \left(\frac{B\dot L}{L} - \frac{\dot B}{2}\right)|y|^2 \\ -i\frac{\dot X}{L}\cdot\nabla - i\frac{\dot L}{L}y\cdot\nabla \end{array}\right]v(t,y). \tag{2.8}$$

Finally, the nonlinearity satisfies

$$|u|^2 u = M_{P(t)}(x)\frac{1}{L^2}|v|^2 v.$$

Gathering the previous calculation yields (2.4). □

Once we have this equation, we are free to choose the parameters as we wish. Until now, the parameters have been completely free, with no condition whatsoever on them or their time derivatives. The main idea is to now choose conditions so that (2.4) becomes an equation on v in variables (t, y) that is "easy" to solve.

2.3. *Two-dimensional solitons*

From now on we assume $d = 2$. Let Q a soliton associated to an eigenvalue ω, i.e. any solution to (1.4):

$$-\Delta_z Q(z) + |z|^2 Q(z) + \mu|Q(z)|^2 Q(z) = \omega Q(z),$$

and define $v(t, y) := Q(y)$. Note that for $\mu = 0$, the equation (1.1) is the linear Harmonic oscillator, and the previous PDE is solved by Hermite functions. We can also construct solutions for $|\mu| \ll 1$ close to any Hermite function by perturbation, see [26]. We have immediately that

$$i\partial_t v = 0$$

and

$$\frac{1}{L^2}\Delta_y v(y) - \frac{|y|^2}{L^2}v(y) - \frac{\mu}{L^2}|v(y)|^2 v(y) = -\frac{\omega}{L^2}v(y).$$

Therefore, with such v, we get

$$(i\partial_t u + \Delta_x u - |x|^2 u - \mu|u|^2 u)(t, x) = M_{P(t)}(x)$$
$$\times \begin{bmatrix} i\left(-\frac{\dot{L}}{L} + \frac{B}{L^2}d\right) - \dot{\gamma} + \beta \cdot \dot{X} - |X|^2 - |\beta|^2 - \frac{\omega}{L^2} \\ + \left(-L\dot{\beta} + \frac{B}{L}\dot{X} - 2\frac{B}{L}\beta - 2LX\right) \cdot y \\ + \left(\frac{B\dot{L}}{L} - \frac{\dot{B}}{2} - \frac{B^2}{L^2} - L^2 + \frac{1}{L^2}\right)|y|^2 \\ +i\left(2\frac{\beta}{L} - \frac{\dot{X}}{L}\right) \cdot \nabla + i\left(2\frac{B}{L^2} - \frac{\dot{L}}{L}\right) y \cdot \nabla \end{bmatrix} v(t, y(t, x)). \quad (2.9)$$

We then choose the time derivatives of the modulation parameters as follows:

$$\left|\begin{array}{l} \dfrac{\dot L}{L} - 2\dfrac{B}{L^2} = 0 \\ -\dot\gamma + \beta\cdot\dot X - |X|^2 - |\beta|^2 - \dfrac{\omega}{L^2} = 0 \\ -L\dot\beta + \dfrac{B}{L}\dot X - 2\dfrac{B}{L}\beta - 2LX = 0 \\ \dfrac{B\dot L}{L} - \dfrac{\dot B}{2} - \dfrac{B^2}{L^2} - L^2 + \dfrac{1}{L^2} = 0 \\ 2\dfrac{\beta}{L} - \dfrac{\dot X}{L} = 0, \end{array}\right. \iff \left|\begin{array}{l} \dot L = 2\dfrac{B}{L} \\ \dot\gamma = |\beta|^2 - |X|^2 - \dfrac{\omega}{L^2} \\ \dot\beta = -2X \\ \dot B = 2\dfrac{B^2+1}{L^2} - 2L^2 \\ \dot X = 2\beta. \end{array}\right. \tag{2.10}$$

Finally, we obtain a solution of the equation (1.1) of the form

$$u(t,x) = \frac{1}{L} e^{i\gamma + iL\beta\cdot y + i\frac{B}{2}|y|^2} Q(y),$$

where $y = \frac{x - X(t)}{L(t)}$ and the parameters $P(t)$ evolve according to the previous set of ordinary differential equations.

Remark 2. In dimension $d \neq 2$, the equation implies that $B = 0$ and $L = 1$, and the solitons are only rotating. The dynamics is however again richer when the nonlinearity $|u|^2 u$ is replaced by $|u|^{2\sigma} u$ where $\sigma = 2/d$ is again L^2 critical in dimension d (with a prefactor $L^{-\frac{d}{2}}$ instead of L^{-1} in the formula (2.3)).

2.4. *Modulation equations*

In this section we want to find explicit expressions for the modulation parameters solving (2.10). Note that the couples (X, β) and (L, B) can be written as Hamiltonian systems:

$$\left|\begin{array}{l} \dot X = \partial_\beta \mathcal{R}(X,\beta) \\ \dot\beta = -\partial_X \mathcal{R}(X,\beta) \end{array}\right. \quad \text{and} \quad \left|\begin{array}{l} \dot L = L\partial_B \mathcal{E}(L,B) \\ \dot B = -L\partial_L \mathcal{E}(L,B), \end{array}\right.$$

where

$$\mathcal{R}(X,\beta) = |X|^2 + |\beta|^2 \quad \text{and} \quad \mathcal{E}(L,B) = \frac{B^2+1}{L^2} + L^2.$$

The couple (L, B) solves a non-canonical Hamiltonian system, so we write $k = \log L$ and obtain

$$\dot k = \partial_B \mathcal{E}(k,B), \quad \dot B = -\partial_k \mathcal{E}(k,B),$$

with $\mathcal{E}(k,B) = (B^2+1)e^{-2k} + e^{2k}$. Finally, the canonical Hamiltonian system is

$$\left| \begin{aligned} \dot{X} &= \partial_\beta \mathcal{H} \\ \dot{\beta} &= -\partial_X \mathcal{H} \\ \dot{k} &= \partial_B \mathcal{H} \\ \dot{B} &= -\partial_k \mathcal{H}, \end{aligned} \right.$$

where the Hamiltonian $\mathcal{H}$ is given by

$$\mathcal{H}(X,\beta,k,B) = \mathcal{R}(X,\beta) + \mathcal{E}(k,B) = |X|^2 + |\beta|^2 + (B^2+1)e^{-2k} + e^{2k}.$$

Let us recall that a general Hamiltonian system can be written under the form

$$\dot{z} = J^{-1}\nabla H(z), \quad J = \begin{pmatrix} 0 & I \\ -I & 0 \end{pmatrix}, \tag{2.11}$$

where $z = (p,q) \in \mathbb{R}^{2N}$, I is the identity matrix of size N, and $H : \mathbb{R}^N \to \mathbb{R}$ a real function. The Poisson bracket between two functions is $\{K,H\} = \nabla K^T J \nabla H$, so that we have for any function $K(z(t))$ where $z(t)$ solves (2.11):

$$\frac{d}{dt}K(z(t)) = \{H,K\}(z(t)).$$

If there exists N functions $I_\ell(z)$, $\ell = 1,\ldots,N$ that are in involution, *i.e.* $\{I_\ell, I_k\} = 0$, $k \neq \ell$, and such that the gradient $\nabla I_\ell(z)$ are independent everywhere for $\ell = 1,\ldots,N$, then the (2.11) is a completely integrable system (see for instance [30, 31] and the references therein). In this case, we can show that there exist *action-angle* variables, *i.e.* new symplectic variables $\tilde{z} = (a,\theta) = \Phi(p,q)$ such that the system (2.11) is written

$$\dot{\tilde{z}} = J^{-1}\nabla \tilde{H}(\tilde{z}),$$

but with a function $\tilde{H}(\tilde{z}) = K(a)$ depending only on the *action* variable a. This shows that the equation in these new variables can be integrated exactly:

$$\left| \begin{aligned} \dot{a} &= -\partial_\theta K(a) = 0 \\ \dot{\theta} &= \partial_a K(a) = \omega(a) \end{aligned} \right. \quad \Rightarrow \quad \left| \begin{aligned} a(t) &= a(0) \\ \theta(t) &= \theta(0) + t\omega(a_0). \end{aligned} \right.$$

In our case, we can immediately see that the Hamiltonians $\mathcal{E}$ and $\mathcal{R}$ are in involution. Hence the system is completely integrable, and we can calculate explicitly the corresponding action-angle variables:

Lemma 3 (Action-angle variables and exact update formulas). *There exists a symplectic change of variable* $(X, B, k, \beta) \mapsto (h, a, \xi, \theta) \in \mathbb{R} \times \mathbb{R}^d \times [0, 2\pi] \times [0, 2\pi]^d$*, such that the Hamiltonian in these variables is given by*

$$E(h, a, \xi, \theta) = 4h + 2a, \tag{2.12}$$

so that the flow in variable (h, a, ξ, θ) *is given by*

$$\left| \begin{array}{l} a(t) = a(0), \\ \theta(t) = \theta(0) + 2t, \end{array} \right. \quad \textit{and} \quad \left| \begin{array}{l} h(t) = h(0), \\ \xi(t) = \xi(0) + 4t. \end{array} \right. \tag{2.13}$$

Moreover, solutions to the dynamical system (2.10) *are given by the explicit formulas:*

$$\begin{aligned} e^{2k(t)} &= L(t)^2 = 2h(t) - \cos(\xi(t))\sqrt{4h(t)^2 - 1}, \\ B(t) &= \sin(\xi(t))\sqrt{4h(t)^2 - 1}, \\ X_n(t) &= \sin(\theta_n(t))\sqrt{2a_n(t)}, \quad n = 1, \dots, d, \\ \beta_n(t) &= \cos(\theta_n(t))\sqrt{2a_n(t)}, \quad n = 1, \dots, d, \\ \gamma(t) &= \sum_{\ell=1}^{d} \frac{a_\ell}{2} \left[\sin(2\theta_\ell(t)) - \sin(2\theta_\ell(0))\right] \\ &\quad + \frac{\omega}{2} \arctan\left(\left(2h(0) + \sqrt{4h(0)^2 - 1}\right) \tan\left(\frac{\xi(t)}{2}\right)\right) \\ &\quad - \frac{\omega}{2} \arctan\left(\left(2h(0) + \sqrt{4h(0)^2 - 1}\right) \tan\left(\frac{\xi(0)}{2}\right)\right) + m_t \frac{\pi}{2}, \end{aligned} \tag{2.14}$$

where, if $m_0 \in \mathbb{Z}$ *is such that* $\frac{\xi(0)}{2} \in m_0\pi + \left[-\frac{\pi}{2}, \frac{\pi}{2}\right]$*, then* $m_t \in \mathbb{Z}$ *is defined by* $\frac{\xi(t)}{2} \in (m_0 - m_t)\pi + \left[-\frac{\pi}{2}, \frac{\pi}{2}\right]$.

Proof. For the (X, β) part, it suffices to check that the change of variable $(X, \beta) \mapsto (a, \theta)$ defined by

$$X_n = \sqrt{2a_n}\sin(\theta_n) \quad \text{and} \quad \beta_n = \sqrt{2a_n}\cos(\theta_n) \qquad n = 1, \dots, d,$$

is symplectic and that

$$X_n = \sqrt{2a_n(0)}\sin(2t + \theta_n(0)) \quad \text{and} \quad \beta_n = \sqrt{2a_n(0)}\cos(2t + \theta_n(0))$$

are solutions. Thus,

$$a_n(t) = a_n(0) \qquad \text{and} \qquad \theta_n(t) = \theta_n(0) + 2t. \tag{2.15}$$

For the (k, B) part we use the method of generating functions, described e.g. in [31, Sect. VI.5]. We can express B in terms of k and the Hamiltonian $\mathcal{E}$, so that on the set $\{B > 0\}$ we have:

$$B = \sqrt{e^{2k}\mathcal{E} - e^{4k} - 1}. \tag{2.16}$$

This equality holds for $e^{2k} \in [e^{2k_0}, e^{2k_1}]$, where e^{2k_0}, e^{2k_1} are the reals roots of the polynomial $-z^2 + \mathcal{E}z - 1$,

$$e^{2k_0} = \frac{1}{2}\left(\mathcal{E} - \sqrt{\mathcal{E}^2 - 4}\right), \quad e^{2k_1} = \frac{1}{2}\left(\mathcal{E} + \sqrt{\mathcal{E}^2 - 4}\right). \tag{2.17}$$

In order to obtain a symplectic change of variables, we look for a function $S(k, \mathcal{E})$ such that

$$B = -\frac{\partial S}{\partial k}(k, \mathcal{E}).$$

We easily obtain $S(k, \mathcal{E})$, by integrating on $[k_0, k]$:

$$S(k, \mathcal{E}) = -\int_{k_0}^{k} \sqrt{e^{2z}\mathcal{E} - e^{4z} - 1}dz.$$

The variable ϕ which makes the mapping $(B, k) \mapsto (\phi, \mathcal{E})$ symplectic is defined by

$$\phi = \frac{\partial S}{\partial \mathcal{E}}(k, \mathcal{E}) = -\frac{1}{2}\int_{k_0}^{k} \frac{e^{2z}}{\sqrt{e^{2z}\mathcal{E} - e^{4z} - 1}}dz.$$

We then have

$$\frac{\mathrm{d}\phi}{\mathrm{d}t} = -\frac{1}{2}\frac{e^{2k}\dot{k}}{\sqrt{e^{2k}\mathcal{E} - e^{4k} - 1}} = -\frac{1}{2}\frac{e^{2k}\partial_B \mathcal{E}}{B} = -\frac{1}{2}\frac{e^{-2k}2Be^{2k}}{B} = -1.$$

We now proceed to obtaining an explicit expression for ϕ:

$$\phi = -\frac{1}{2}\int_{k_0}^{k} \frac{e^{2z}}{\sqrt{e^{2z}\mathcal{E} - e^{4z} - 1}}dz = -\frac{1}{4}\int_{e^{2k_0}}^{e^{2k}} \frac{1}{\sqrt{\mathcal{E}u - u^2 - 1}}du$$

$$= -\frac{1}{4\sqrt{\mathcal{E}^2/4 - 1}}\int_{e^{2k_0}}^{e^{2k}} \frac{1}{\sqrt{1 - \left(\frac{u - \mathcal{E}/2}{\sqrt{\mathcal{E}^2/4 - 1}}\right)^2}}du = -\frac{1}{4}\int_{\frac{e^{2k_0} - \frac{\mathcal{E}}{2}}{\sqrt{\mathcal{E}^2/4 - 1}}}^{\frac{e^{2k} - \frac{\mathcal{E}}{2}}{\sqrt{\mathcal{E}^2/4 - 1}}} \frac{1}{\sqrt{1 - r^2}}dr.$$

Recall the definition (2.17) of k_0, which yields

$$e^{2k_0} - \mathcal{E}/2 = -\sqrt{\mathcal{E}^2/4 - 1}.$$

Therefore,

$$\phi = -\frac{1}{4}\int_{-1}^{\frac{e^{2k} - \mathcal{E}/2}{\sqrt{\mathcal{E}^2/4 - 1}}} \frac{1}{\sqrt{1 - r^2}}dr = -\frac{1}{4}\left(\arcsin\left(\frac{e^{2k} - \mathcal{E}/2}{\sqrt{\mathcal{E}^2/4 - 1}}\right) + \frac{\pi}{2}\right)$$

$$= -\frac{1}{4}\arcsin\left(\frac{e^{2k} - \mathcal{E}/2}{\sqrt{\mathcal{E}^2/4 - 1}}\right) - \frac{\pi}{8} \in \left[-\frac{\pi}{4}, 0\right].$$

We want the angle variable to lie in $[-\pi, \pi]$ so the above expression describes an eighth of a period. But we are only considering the set $\{B > 0\}$, thus the angle ξ we are looking for must lie only in $[0, \pi]$. Hence we set $(\xi, h) = (-4\phi, \mathcal{E}/4)$ and let the Hamiltonian $\mathcal{E}(\xi, h) = 4h$ with a slight abuse of notation. It is then clear that $\frac{\mathrm{d}h}{\mathrm{d}t} = 0$ and $\frac{\mathrm{d}\xi}{\mathrm{d}t} = 4$. Moreover,

$$\xi = \arcsin\left(\frac{e^{2k} - \mathcal{E}/2}{\sqrt{\mathcal{E}^2/4 - 1}}\right) + \frac{\pi}{2} \in [0, \pi], \tag{2.18}$$

and hence

$$\frac{e^{2k} - \mathcal{E}/2}{\sqrt{\mathcal{E}^2/4 - 1}} = \sin\left(\xi - \frac{\pi}{2}\right) = -\cos(\xi).$$

We obtain

$$\begin{aligned} e^{2k} = L^2 &= \frac{\mathcal{E}}{2} - \cos(\xi)\sqrt{\mathcal{E}^2/4 - 1} = 2h - \cos(\xi)\sqrt{4h^2 - 1} \\ &= 2h\left(1 - \cos(\xi)\sqrt{1 - 1/(4h^2)}\right). \end{aligned}$$

With this formula, we have

$$0 < L^2 < 4h = \mathcal{E},$$

and (2.16) becomes

$$\begin{aligned} B &= \sqrt{\mathcal{E}e^{2k} - e^{4k} - 1} = \sqrt{4he^{2k} - (e^{2k})^2 - 1} = \sqrt{(4h^2 - 1)\sin^2(\xi)} \\ &= \sin(\xi)\sqrt{4h^2 - 1}, \end{aligned}$$

where the last equality holds for $\xi \in [0, \pi]$.

We can now integrate the equation on γ, using direct integration:

$$\begin{aligned} \int_0^t \left[|\beta(\tau)|^2 - |X(\tau)|^2\right] d\tau &= \int_0^t \left\{\sum_{\ell=1}^d 2a_\ell \cos(\theta_\ell(\tau))^2 - \sum_{\ell=1}^d 2a_\ell \sin(\theta_\ell(\tau))^2\right\} d\tau \\ &= \int_0^t 2\sum_{\ell=1}^d a_\ell \left(\cos(\theta_\ell(\tau))^2 - \sin(\theta_\ell(\tau))^2\right) d\tau \\ &= \int_0^t \sum_{\ell=1}^d 2a_\ell \cos(2\theta_\ell(\tau)) d\tau \\ &= \sum_{\ell=1}^d \frac{a_\ell}{2} \left[\sin(2\theta_\ell(t)) - \sin(2\theta_\ell(0))\right], \end{aligned}$$

where the last equality has been obtained using (2.15). We compute

$$\int_0^t \frac{1}{L(\tau)^2} dt = \int_0^t \frac{1}{\underbrace{2h(0)}_{=:c_1} - \underbrace{\sqrt{4h(0)^2-1}}_{=:c_2} \cos(\xi(0)+4r)} dr$$
$$= \int_0^t \frac{1}{c_1 - c_2\cos(\xi(0)+4r)} dr = \frac{1}{4}\int_{\xi(0)}^{\xi(0)+4t} \frac{1}{c_1 - c_2\cos(\tau)} d\tau.$$

Recall the following trigonometric identity:

$$\cos(2\tau) = \frac{1-\tan(\tau)^2}{1+\tan(\tau)^2}, \quad \tau \in \mathbb{R},$$

hence

$$\int_0^t \frac{1}{c_1 - c_2\cos(\xi(0)-4\tau)} d\tau$$
$$= \frac{1}{4}\int_{\xi(0)}^{\xi(0)+4t} \frac{1}{c_1 - c_2 \frac{1-\tan(\tau/2)^2}{1+\tan(\tau/2)^2}} d\tau$$
$$= \frac{1}{4}\int_{\xi(0)}^{\xi(0)+4t} \frac{1+\tan(\tau/2)^2}{c_1(1+\tan(\tau/2)^2) - c_2(1-\tan(\tau/2)^2)} d\tau$$
$$= \frac{1}{4}\int_{\xi(0)}^{\xi(0)+4t} \frac{1+\tan(\tau/2)^2}{(c_1+c_2)\tan(\tau/2)^2 + c_1 - c_2} d\tau$$
$$= \frac{1}{4(c_1-c_2)}\int_{\xi(0)}^{\xi(0)+4t} \frac{1+\tan(\tau/2)^2}{\frac{c_1+c_2}{c_1-c_2}\tan(\tau/2)^2 + 1} d\tau$$
$$= \frac{1}{2(c_1-c_2)}\int_{\frac{\xi(0)}{2}}^{\frac{\xi(0)}{2}+2t} \frac{1+\tan(\tau)^2}{\frac{c_1+c_2}{c_1-c_2}\tan(\tau)^2 + 1} d\tau$$
$$= \frac{1}{2(c_1-c_2)}\int_{\frac{\xi(0)}{2}}^{\frac{\xi(0)}{2}+2t} \frac{\frac{d}{d\tau}(\tan(\tau))}{\frac{c_1+c_2}{c_1-c_2}\tan(\tau)^2 + 1} d\tau$$
$$= \frac{1}{2(c_1-c_2)}\frac{1}{\sqrt{\frac{c_1+c_2}{c_1-c_2}}}\int_{\frac{\xi(0)}{2}}^{\frac{\xi(0)}{2}+2t} \frac{\frac{d}{d\tau}\left(\sqrt{\frac{c_1+c_2}{c_1-c_2}}\tan(\tau)\right)}{\left[\sqrt{\frac{c_1+c_2}{c_1-c_2}}\tan(\tau)\right]^2 + 1} d\tau.$$

Moreover, $(c_1-c_2)(c_1+c_2) = c_1^2 - c_2^2 = (2h)^2 - (4h^2-1) = 1$ and $c_1 - c_2 > 0$, thus $\sqrt{\frac{c_1+c_2}{c_1-c_2}} = (c_1+c_2)$ and

$$\int_0^t \frac{1}{L(\tau)^2} d\tau = \frac{1}{2}\int_{\frac{\xi(0)}{2}}^{\frac{\xi(0)}{2}+2t} \frac{\frac{d}{d\tau}((c_1+c_2)\tan(\tau))}{((c_1+c_2)\tan(\tau))^2+1} d\tau.$$

Now let $m_0 \in \mathbb{Z}$ such that $\frac{\xi(0)}{2} \in m_0\pi + \left(-\frac{\pi}{2}, \frac{\pi}{2}\right]$, and $m_t \in \mathbb{Z}$ such that $\frac{\xi(t)}{2} \in (m_0 - m_t)\pi + \left(-\frac{\pi}{2}, \frac{\pi}{2}\right]$. We recall that $\xi(t) = \xi(0) + 4t$. Then

$$\begin{aligned}
\int_0^t \frac{1}{L(\tau)^2} d\tau &= \frac{1}{2} \int_{\frac{\xi(0)}{2}}^{\frac{\xi(0)}{2}+2t} \underbrace{\frac{\frac{d}{d\tau}\left((c_1+c_2)\tan(\tau)\right)}{\left((c_1+c_2)\tan(\tau)\right)^2+1}}_{=:f(\tau)} d\tau \\
&= \frac{1}{2} \int_{\frac{\xi(0)}{2}}^{m_0\pi+\frac{\pi}{2}} f(\tau)d\tau + \frac{1}{2} \int_{(m_0+1)\pi-\frac{\pi}{2}}^{(m_0+1)\pi+\frac{\pi}{2}} f(\tau)d\tau \\
&\quad + \cdots + \frac{1}{2} \int_{(m_0+m_t)\pi-\frac{\pi}{2}}^{\frac{\xi(0)}{2}+2t} f(\tau)d\tau.
\end{aligned}$$

For $m \in \mathbb{Z}$, we have

$$\begin{aligned}
\int_{m\pi-\frac{\pi}{2}}^{m\pi+\frac{\pi}{2}} f(\tau)d\tau &= \left[\arctan\left((c_1+c_2)\tan(\tau)\right)\right]_{m\pi-\frac{\pi}{2}}^{m\pi+\frac{\pi}{2}} \\
&= \left[\arctan\left((c_1+c_2)\tan(\tau)\right)\right]_{-\frac{\pi}{2}}^{\frac{\pi}{2}} = \pi.
\end{aligned}$$

Now write $\widetilde{\frac{\xi(0)}{2}} := \frac{\xi(0)}{2} - m_0\pi \in \left(-\frac{\pi}{2}, \frac{\pi}{2}\right]$, and $\widetilde{\frac{\xi(\tau)}{2}} := \frac{\xi(\tau)}{2} - (m_0 - m_t)\pi \in \left(-\frac{\pi}{2}, \frac{\pi}{2}\right]$. Then,

$$\begin{aligned}
&\int_0^t \frac{1}{L(\tau)^2} d\tau \\
&= \frac{1}{2}(m_t - 1)\pi + \frac{1}{2} \int_{\frac{\xi(0)}{2}}^{m_0\pi+\frac{\pi}{2}} f(\tau)d\tau + \frac{1}{2} \int_{(m_0+m_t)\pi-\frac{\pi}{2}}^{\frac{\xi(0)}{2}+2t} f(\tau)d\tau \\
&= (m_t - 1)\frac{\pi}{2} + \frac{1}{2} \int_{\widetilde{\frac{\xi(0)}{2}}}^{\frac{\pi}{2}} f(\tau)d\tau + \frac{1}{2} \int_{-\frac{\pi}{2}}^{\widetilde{\frac{\xi(t)}{2}}} f(\tau)d\tau \\
&= (m_t - 1)\frac{\pi}{2} + \frac{1}{2}\left[\arctan\left((c_1+c_2)\tan(\tau)\right)\right]_{\widetilde{\frac{\xi(0)}{2}}}^{\frac{\pi}{2}} + \frac{1}{2}\left[\arctan\left((c_1+c_2)\tan(\tau)\right)\right]_{-\frac{\pi}{2}}^{\widetilde{\frac{\xi(t)}{2}}} \\
&= (m_t - 1)\frac{\pi}{2} + \frac{\pi}{2} - \frac{1}{2}\arctan\left((c_1+c_2)\tan\left(\widetilde{\frac{\xi(0)}{2}}\right)\right) \\
&\quad + \arctan\left((c_1+c_2)\tan\left(\widetilde{\frac{\xi(t)}{2}}\right)\right) + \frac{\pi}{2} \\
&= m_t\frac{\pi}{2} + \frac{1}{2}\arctan\left((c_1+c_2)\tan\left(\widetilde{\frac{\xi(t)}{2}}\right)\right) - \frac{1}{2}\arctan\left((c_1+c_2)\tan\left(\widetilde{\frac{\xi(0)}{2}}\right)\right) \\
&= m_t\frac{\pi}{2} + \frac{1}{2}\arctan\left((c_1+c_2)\tan\left(\frac{\xi(t)}{2}\right)\right) - \frac{1}{2}\arctan\left((c_1+c_2)\tan\left(\frac{\xi(0)}{2}\right)\right)
\end{aligned}$$

Hence

$$\begin{aligned}\gamma(t) &= \int_0^t \left[|\beta(\tau)|^2 - |X(\tau)|^2 + \frac{\omega}{L(\tau)^2}\right] d\tau + \gamma(0)\\ &= \gamma(0) + \sum_{\ell=1}^{d} \frac{a_\ell}{2}\left[\sin(2\theta_\ell(t)) - \sin(2\theta_\ell(0))\right]\\ &\quad + \frac{\omega}{2}\arctan\left((c_1+c_2)\tan\left(\frac{\xi(t)}{2}\right)\right)\\ &\quad - \frac{\omega}{2}\arctan\left((c_1+c_2)\tan\left(\frac{\xi(0)}{2}\right)\right) + m_t\frac{\pi}{2}.\end{aligned}$$

□

Lemma 4 (Action-angle variables from the parameters). *The change of variables $(L, B, X, \beta) \mapsto (h, a, \xi, \theta)$ is explicit, and at time $t = 0$ we have*

$$\begin{aligned}a_n(0) &= \frac{1}{2}\left(X_n(0)^2 + \beta_n(0)^2\right), \quad n = 1, \ldots, d,\\ \theta_n(0) &= \arctan\left(\frac{X_n(0)}{\beta_n(0)}\right), \quad n = 1, \ldots, d,\\ h(0) &= \frac{L(0)^4 + 1 + B(0)^2}{4L(0)^2},\\ \xi(0) &= \arctan\left(\frac{B(0)}{2h(0) - L(0)^2}\right),\end{aligned} \tag{2.19}$$

whenever $\theta_n(0)$ and $\xi(0)$ are well-defined. When any one of them is ill-defined – which happens when $X_n(0) = \beta_n(0) = 0, n \in \{1, \ldots, d\}$ or when $L(0) = 1$ and $B(0) = 0$ – any value can be taken and the time evolution of $A(t)$, $L(t)$, $B(t)$, $X(t)$, $\beta(t)$ and $\gamma(t)$ will not depend on the value. Moreover, in the cases where $a_n(0) = 0$, $n \in \{1, \ldots, d\}$ or $h(0) = \frac{1}{2}$, the formula (2.19) for $\theta_n(0)$, $n \in \{1, \ldots, d\}$ or $\xi(0)$ are ill-defined, but any value can be taken as a substitution and this will not affect the behavior of the mappings $t \mapsto \gamma(t)$ and $t \mapsto s(t)$.

Proof. The proof of this lemma consists in inverting the expressions (2.14) using algebraic manipulations. □

Let us sum up this section: we have shown that, if the initial condition U_0 to the nonlinear Schrödinger equation can be written as a modulated soliton $U_0 = \mathcal{M}_{P_0}Q$, then it is possible to write the time-dependent solution $u(t)$ as a modulated soliton. In other words, $u(t, x) = (\mathcal{M}_{P(t)}Q)(t, x)$.

Since the time-dependence is now only through the parameters P, we write $u(t,x) = (\mathcal{M}_{P(t)}Q)(x)$. The time-dependent modulation parameters involved in $\mathcal{M}_{P(t)}$ are known exactly for all times t by Lemma 3.

3. Two or more solitons

3.1. *Modulated solitons manifold*

Assume now that the initial condition U_0 is a sum of N solitons

$$U_0 = \sum_{n=1}^{N} u^n := \sum_{n=1}^{N} \mathcal{M}_{P_0^n} v^n,$$

where P_0^n are the initial modulation parameters associated to the n-th function v^n. We recall that the superscript is not meant as an exponent here.

Even though the computation of the nonlinearity $u|u|^2$ involves the interaction of all triplets of solitons, the most important interaction for each soliton will be the self-interaction. In this section we will show that the nonlinear Schrödinger equation involving u can be split into N nonlinear Schrödinger equations involving u^n, up to some generally less important interaction terms. We can then solve each nonlinear Schrödinger equation on u^n using Section 2.2, and the remaining interaction terms can be approximated numerically using a procedure based on the Dirac-Frenkel principle.

We focus now on (1.2) in the case of modulating soliton. It means that we search an approximation

$$u(t,x) \approx \sum_{j=1}^{N} u^j(t,x) := \sum_{j=1}^{N} \frac{\exp\left\{i\gamma^j + iL^j\beta^j \cdot y^j + i\frac{B^j}{2}|y^j|^2\right\}}{(L^j)^{d/2}} Q_j(y^j), \tag{3.1}$$

where all the parameters are time-dependent, and where Q^j are solitons solutions of (1.4) with possible different frequencies ω_j. For simplicity, we will assume in the following that all solitons are associated to the same eigenvalue ω, i.e. that we have $Q^j = Q$ for all j, where Q is a given solution of (1.4).

Remark 5. We do not address here the problem of general approximation of any function by a sum of modulated solitons. Here, we assume that the initial datum is under this form, which is relevant in many physical applications, see for instance [9, 19] and the references therein.

The numerical method we are looking for aims at approximating the time evolution of all the parameters while keeping the modulation ansatz. This naturally calls for the Dirac-Frenkel principle. Let $\mathcal{V}$ be the modulation manifold defined by

$$\mathcal{V} := \left\{ u \in L^2(\mathbb{R}^d) \middle| \begin{array}{c} u(x) = \sum_{j=1}^{N} (\mathcal{M}_{P^j} Q)(x) \in L^2(\mathbb{R}^d) \\ P^j = (\gamma^j, X^j, \beta^j, B^j, L^j) \in \mathbb{R}^{2d+2} \times \mathbb{R}_+, \; 1 \le j \le N \end{array} \right\}.$$

Let $\mathcal{T}_{u(t)}\mathcal{V}$ be the tangent space at $u(t)$ of $\mathcal{V}$ viewed as a real submanifold of L^2 (itself viewed as a real Hilbert space associated with the scalar product $\langle \cdot, \cdot \rangle$ defined by (2.1)). Note that any point on the manifold $\mathcal{V}$ can be parameterized by the coordinate vector

$$P := \{P^1, \dots, P^N\},$$

and if we write $u(x) = u_P(x)$, the projection of (1.1) on the manifold can be written (through the Dirac-Frenkel principle) for $u(t,x) = u_{P(t)}(x)$:

$$\forall f \in T_{u(t)}\mathcal{V} \quad \langle f, \partial_t u - i\Delta u + i|x|^2 u + i\mu |u|^2 u \rangle = 0.$$

By taking a local basis of $\mathcal{T}_{u(t)}\mathcal{V}$, this equation can be recast as a nonlinear system of ordinary differential equation on P, of the form

$$B(P)\dot{P} = F(P),$$

where $B(P)$ is a matrix and $F(P)$ a vector field. The exact expression of P will be revealed later, but first we propose a natural splitting of the vector field which takes advantage of the structure of the soliton.

3.2. *Splitting of the equation*

The vector field $F(P)$ can be decomposed according to the structure of $u = \sum_j u^j$. Let us compute $|u|^2 u$:

$$\left|\sum u^j\right|^2 \sum u^j = \left(\sum_j u^j\right)\left(\sum_k \overline{u^k}\right)\sum_l u^l = \sum_{j,k,l} u^j \overline{u^k} u^l$$
$$= \sum_{j=1}^{N} |u^j|^2 u^j + \sum_{j=1}^{N} \sum_{(k,l)\neq(j,j)} u^j \overline{u^k} u^l.$$

Hence we can write

$$B(P)\dot{P} = F(P) = F_1(P) + F_2(P), \tag{3.2}$$

where F_1 corresponds to the part

$$i\partial_t u = -\Delta_x \sum_{j=1}^N u^j + |x|^2 \sum_{j=1}^N u^j + \mu \sum_{j=1}^N |u^j|^2 u^j, \tag{F1}$$

which is the *soliton* part, and F_2 the remainder term corresponding to the equation

$$i\partial_t u = \mu \sum_j \sum_{(k,l)\neq(j,j)} u^j \overline{u^k} u^l \tag{F2}$$

projected onto the manifold $\mathcal{V}$, which is the *interactions* part.

Note that when the solitons are rather localized and their centers X^k are all far from each other, we expect the interaction terms to be small: $|\sum_{j=1}^N \sum_{(k,l)\neq(j,j)} u^j\overline{u^k}u^l| \ll 1$ (typically exponentially decreasing with respect to the distance between the solitons).

Now, the content of Section 2.2 precisely shows that the (F1) equation can be solved exactly on the manifold $\mathcal{V}$ by choosing $u^j = \mathcal{M}_{P^j}Q$, separately solving the equation

$$\forall j = 1, \dots, N, \quad i\partial_t u^j + \Delta_x u^j - |x|^2 u^j = \mu |u^j|^2 u^j, \tag{3.3}$$

through the modulation equations described in Section 2.4.

Hence we only need to project the equation (F2) onto the manifold $\mathcal{V}$ to obtain a natural splitting of the equation (3.2). The Dirac-Frenkel principle applied to (F2) consists in looking for solutions to

Find $\partial_t u(t) \in \mathcal{T}_{u(t)}\mathcal{V}$, such that

$$\left\langle f, i\partial_t \sum_{j=1}^N u^j - \mu \sum_{j=1}^N \sum_{\substack{k,l=1\\(k,l)\neq(j,j)}}^N u^j\overline{u^k}u^l \right\rangle = 0, \quad \forall f \in \mathcal{T}_{u(t)}\mathcal{V}. \tag{3.4}$$

Let $S_{u(t)}$ be a basis of $\mathcal{T}_{u(t)}\mathcal{V}$, then (3.4) is equivalent to

Find $\partial_t u(t) \in \mathcal{T}_{u(t)}\mathcal{V}$, such that

$$\left\langle f, i\partial_t \sum_{j=1}^N u^j \right\rangle = \mu \left\langle f, \sum_{j=1}^N \sum_{\substack{k,l=1\\(k,l)\neq(j,j)}}^N u^j\overline{u^k}u^l \right\rangle, \quad \forall f \in S_{u(t)}. \tag{3.5}$$

A family (which may happen to be linearly dependent) spanning the tangent space $\mathcal{T}_{u(t)}\mathcal{V}$ is given by

$$\begin{aligned} S_{u(t)} &= \left\{ \begin{matrix} e^{i\Gamma^j(y^j)}Q(y^j), (y_1^j)e^{i\Gamma^j(y^j)}Q(y^j), \dots, (y_d^j)e^{i\Gamma^j(y^j)}Q(y^j), \\ ie^{i\Gamma^j(y^j)}\partial_{y_1}Q(y^j), \dots, ie^{i\Gamma^j(y^j)}\partial_{y_d}Q(y^j), |y^j|^2 e^{i\Gamma^j(y^j)}Q(y^j), \\ ie^{i\Gamma^j(y^j)}\left(1 + y\cdot\nabla\right)Q(y^j) \end{matrix} \right\}_{j=1}^N \\ &=: \left\{ b_0^j, \dots, b_{2d+2}^j : j = 1, \dots, N \right\}, \end{aligned} \tag{3.6}$$

where we defined

$$\Gamma^j(y^j) := \gamma^j + L^j\beta^j \cdot y^j + \frac{B^j}{2}|y^j|^2.$$

This family $S_{u(t)}$ indeed spans the tangent space $\mathcal{T}_{u(t)}\mathcal{V}$ because $\partial_t u$ writes as a linear combination of all the b^p_q, by (2.8). We can rewrite (3.5) as

$$\left|\begin{array}{l}\text{Find } \partial_t u(t) \in \mathcal{T}_{u(t)}\mathcal{V}, \text{ such that for } p \in [\![1, N]\!], q \in [\![0, 2d+2]\!],\\ \left\langle b^p_q, i\partial_t \sum_{j=1}^{N} u^j \right\rangle = \mu \left\langle b^p_q, \sum_{\substack{j,k,l=1\\(k,l)\neq(j,j)}}^{N} u^j\overline{u^k}u^l \right\rangle.\end{array}\right. \tag{3.7}$$

The next step consists in expressing (3.7) as a linear system involving the modulation parameters and their time derivative, which will yield ordinary differential equations that can be integrated numerically.

For $u \in \mathcal{V}$, we can write (2.8) in a more convenient form:

$$i\partial_t u = \sum_{j=1}^{N} \left(b^j_0, \ldots, b^j_{2d+2}\right) \underbrace{\frac{1}{(L^j)^{d/2}} \begin{pmatrix} -\dot{\gamma}^j + \beta^j \cdot \dot{X}^j \\ -L^j\dot{\beta}^j_1 + \frac{B^j}{L^j}\dot{X}^j_1 \\ \vdots \\ -L^j\dot{\beta}^j_d + \frac{B^j}{L^j}\dot{X}^j_d \\ -\frac{\dot{X}^j_1}{L^j} \\ \vdots \\ -\frac{\dot{X}^j_d}{L^j} \\ \frac{B^j\dot{L}^j}{L^j} - \frac{\dot{B}^j}{2} \\ -\frac{\dot{L}^j}{L^j} \end{pmatrix}}_{=:D^j}$$

$$= \left(b^1_0, \ldots, b^1_{2d+2}, b^2_0, \ldots, b^N_{2d+2}\right) \begin{pmatrix} D^1 \\ \vdots \\ D^N \end{pmatrix}.$$

We are now able to rewrite (3.7) under matrix form

$$\mathbf{A}D = S, \tag{3.8}$$

where $\mathbf{A} \in \mathbb{C}^{(2d+3)N\times(2d+3)N}$ is the projection matrix, $D \in \mathbb{R}^{(2d+3)N}$ is a vector containing the modulation parameters and their derivative, and $S \in \mathbb{C}^{(2d+3)N}$ contains the projection of the nonself-interactions. More precisely,

$$D = \begin{pmatrix} D^1 \\ \vdots \\ D^N \end{pmatrix} \tag{3.9}$$

with $D^j \in \mathbb{R}^{2d+3}$,

$$\mathbf{A} = \begin{pmatrix} \langle b_0^1, b_0^1, \rangle & \dots & \langle b_0^1, b_{2d+2}^1 \rangle & \langle b_0^1, b_0^2 \rangle & \dots & \langle b_0^1, b_{2d+2}^N \rangle \\ \vdots & & \vdots & \vdots & & \vdots \\ \langle b_{2d+2}^N, b_0^1, \rangle & \dots & \langle b_{2d+2}^N, b_{2d+2}^1 \rangle & \langle b_{2d+2}^N, b_0^2 \rangle & \dots & \langle b_{2d+2}^N, b_{2d+2}^N \rangle \end{pmatrix}, \tag{3.10}$$

and

$$S = \mu \begin{pmatrix} \left\langle b_0^1, \sum_{j=1}^N \sum^N{}_{\substack{k,l=1 \\ (k,l)\neq(j,j)}} u^j \overline{u^k} u^l \right\rangle \\ \vdots \\ \left\langle b_{2d+2}^N, \sum_{j=1}^N \sum^N{}_{\substack{k,l=1 \\ (k,l)\neq(j,j)}} u^j \overline{u^k} u^l \right\rangle \end{pmatrix}. \tag{3.11}$$

It will be convenient later to write $D^j = \mathbf{C}^j \dot{P}^j$, where

$$P^j = \left(\gamma^j, X_1^j, \dots, X_d^j, \beta_1^j, \dots, \beta_d^j, B^j, L^j\right), \tag{3.12}$$

and

$$\mathbf{C}^j = \begin{pmatrix} -1 & \beta_1^j & \dots & \beta_d^j & 0 & \dots & 0 & 0 & 0 \\ 0 & \frac{B^j}{L} & & 0 & -L^j & & 0 & 0 & 0 \\ \vdots & & \ddots & & & \ddots & & \vdots & \vdots \\ 0 & 0 & & \frac{B^j}{L} & 0 & & -L^j & 0 & 0 \\ 0 & -\frac{1}{L^j} & & 0 & 0 & \dots & 0 & 0 & 0 \\ 0 & & \ddots & & 0 & \dots & 0 & 0 & 0 \\ 0 & 0 & & -\frac{1}{L^j} & 0 & \dots & 0 & 0 & 0 \\ 0 & 0 & \dots & 0 & 0 & \dots & 0 & -\frac{1}{2} & \frac{B^j}{L^j} \\ 0 & 0 & \dots & 0 & 0 & \dots & 0 & 0 & -\frac{1}{L^j} \end{pmatrix} \in \mathbb{C}^{(2d+3)\times(2d+3)}. \tag{3.13}$$

It is readily seen that $\mathbf{C}$ is invertible for $L \neq 0$.

Once the linear system (3.8) is solved, we obtain the solution vector D, from which we can update the modulation parameters. In order to solve numerically the linear system, we rewrite it to use only real matrices. Let $\mathbf{A}_\Re := \Re(\mathbf{A})$, $\mathbf{A}_\Im := \Im(\mathbf{A})$, $D_\Re := \Re(D)$, $D_\Im := \Im(D)$, $S_\Re := \Re(S)$, and $S_\Im := \Im(S)$.

$$\begin{aligned} \mathbf{A}D = S &\iff (\mathbf{A}_\Re + i\mathbf{A}_\Im)(D_\Re + iD_\Im) = S_\Re + iS_\Im \\ &\iff \begin{cases} \mathbf{A}_\Re D_\Re - \mathbf{A}_\Im D_\Im = S_\Re \\ \mathbf{A}_\Im D_\Re + \mathbf{A}_\Re D_\Im = S_\Im \end{cases} \\ &\iff \begin{pmatrix} \mathbf{A}_\Re & -\mathbf{A}_\Im \\ \mathbf{A}_\Im & \mathbf{A}_\Re \end{pmatrix} \begin{pmatrix} D_\Re \\ D_\Im \end{pmatrix} = \begin{pmatrix} S_\Re \\ S_\Im \end{pmatrix}. \end{aligned} \tag{3.14}$$

It is more convenient to solve (3.14) than (3.8), because we only have to deal with real matrices and vectors. Once (3.14) is solved, we obtain (3.9), where each D^j is given by

$$D^j = \begin{pmatrix} -\dot{\gamma}^j + \beta^j \cdot \dot{X}^j \\ -L^j \dot{\beta}_1^j + \frac{B^j}{L^j} \dot{X}_1^j \\ \vdots \\ -L^j \dot{\beta}_d^j + \frac{B^j}{L^j} \dot{X}_d^j \\ -i\frac{\dot{X}_1^j}{L^j} \\ \vdots \\ -i\frac{\dot{X}_d^j}{L^j} \\ \frac{B^j \dot{L}^j}{L^j} - \frac{\dot{B}^j}{2} \\ -i\frac{\dot{L}^j}{L^j} \end{pmatrix} = (L^j)^{d/2}\left(D^j_{\Re} + iD^j_{\Im}\right).$$

Since D^j is a real vector, we get

$$D^j = \mathbf{C}^j \dot{P}^j = (L^j)^{d/2} D^j_{\Re} \quad \Longleftrightarrow \quad \dot{P}^j = (\mathbf{C}^j)^{-1}(L^j)^{d/2} D^j_{\Re}. \tag{3.15}$$

The imaginary part $D^j_{\Im}$ is discarded due to the approximation made by the Dirac-Frenkel principle. The combination of (3.14) and (3.15) corresponds to the F_2 part of the system (3.2). Note that this part is a highly nonlinear system in terms of P, as all the terms in the matrices depend on P. Moreover, the interaction term S defined in (3.11) has to be implemented numerically.

3.3. *Computing interactions numerically*

3.3.1. *Numerical approximation of solitons*

In the ansatz (3.1) we need to be able to compute the soliton Q numerically. To do this, we use the classical Normalized Gradient method (see [32] and the references therein), which gives an approximation of the soliton on a numerical grid. Let us recall that it consists in approximating the solution of (1.4) by a combination of the discretization of the parabolic equation

$$\left| \begin{aligned} &\partial_t \psi = (\Delta - |x|^2)\psi - \mu|\psi|^2\psi, \\ &\psi(\ell\tau) = \psi^{(\ell)}, \end{aligned} \right. \tag{3.16}$$

on a small time interval $[\ell\tau, (\ell+1)\tau]$, which gives a function $\tilde{\psi}^{(\ell)}$, followed by a renormalization step $\tilde{\psi}^{(\ell)} \mapsto \psi^{(\ell+1)} = \frac{\tilde{\psi}^{(\ell)}}{\|\tilde{\psi}^{(\ell)}\|_{L^2}}$, for $\ell = 1, 2, \ldots$. The discretization of the parabolic equation (3.16) is usually done using the implicit Euler scheme, and the global scheme is expected to converge towards the solution of (1.4) of minimal energy and L^2 norm 1 (the ground state, radially symmetric and decreasing with respect to the radius) as proved in [32] in the 1D case. Note that we can easily adapt this algorithm to obtain a soliton of any norm. We can also adapt this algorithm to solitons with specific rotational symmetries to obtain other type of solutions to (1.4).

In the numerical examples below, we choose a finite-size grid, with N nodes along each dimension, and perform the Normalized Gradient method on this grid. We can use a Discrete Fourier transform to have a better computation of the Laplacian than finite differences. We can also seek solitons Q with specific rotational symmetries in 2D and write the equation in polar coordinates, but the grid matching between solitons is slightly more complicated to implement. Once the approximation $\widetilde{Q}$ of the exact soliton Q is obtained on this grid, it is extended by zero outside of the grid and interpolated using periodic B-splines of order 3 to have a continuous representation of the soliton approximation. More details on B-splines can be found in [33–35]. The derivatives of $\widetilde{Q}$ are approximated using a pseudo-spectral method – in this case the Fourier transform, via the Fast Fourier Transform – extended by zero outside the grid, and interpolated using periodic cubic splines.

Note that the Normalized Gradient method can be performed as part of a pre-computation step, and only the values of the soliton on the grid need to be saved. This pre-computation will be done in our numerical experiments to avoid the consequent overhead of computing the soliton values on a fine grid.

3.3.2. *Interaction grids*

In (3.10) and (3.11), we need to compute $L^2(\mathbb{R}^d)$ inner product between solitons centered at possibly different locations. The L^2 inner product $\langle f, g\rangle$ can be approximated using a grid if both functions f and g are known on this grid. If it is not the case – and it is in practice never the case as the translation vector X^j are arbitrary – we have to somehow get an approximation of the two functions on the same grid.

We have chosen to first obtain f and g on a grid, and then perform a linear interpolation of both functions. They are extended by zero outside of their respective grids, and this allows us to compute efficiently the inner product by only approximating the integral on a region that is common to both grids.

3.3.3. *Time integration*

In the integration of the system (3.2), we have seen that we can integrate the $F_1(P)$ part exactly. The interaction part $F_2(P)$ obtained by the Dirac-Frenkel principle are then integrated numerically using the midpoint formula (see for instance [31]).

3.3.4. *Detailed algorithm*

Algorithm 1 Modulation based algorithm with Strang splitting

Require:

- M: number of discretization points for each dimension
- $x_{\min}, x_{\min}$: the grid on which to look for an approximation of the soliton is $[x_{\min}, x_{\max}]^d$
- N: number of modulated solitons in the ansatz
- $L^j, X^j, \beta^j, B^j, \gamma^j$ for $j = 1, \ldots, N$: initial parameters of each modulated soliton
- $\Delta t, T$: stepsize and final time of the simulation
- η: numerical tolerance

Apply the Normalized Gradient method to obtain an approximation of Q on $[x_{\min}, x_{\max}]^d$ using M points for each dimension and a tolerance η.
Interpolate the values on the grid, and extend it by zero outside the grid.
for $t = \Delta t, 2\Delta t, \ldots, T$ **do**
  Update the modulation parameters on a timestep $\frac{\Delta t}{2}$ using Lemma 3
  Use the Dirac-Frenkel equations (3.14)-(3.15) to approximate the time evolution of the modulation parameters, over a timestep Δt
  Update the modulation parameters on a timestep $\frac{\Delta t}{2}$ using Lemma 3
end for

3.4. *Study of error with a single modulated soliton*

We show that the scheme presented above yields a good approximation of the true solution, in the simple case of a single soliton. In this situation, the parameters are given for all times by Lemma 3 since the Dirac-Frenkel principle has right-hand side $S = 0$.

Let us resume our setting: we are looking for a solution

$$u(t,x) = (\mathcal{M}_{P(t)}Q)(x)$$

to (1.1) but in practice, we cannot know Q exactly and we can only approximate it on a grid. For simplicity, let us assume this grid to be defined as follows: let $G = [-\alpha, \alpha]^2$ for some $\alpha > 0$, and write $G_{\Delta x}$ its discretization using a uniform stepsize $\Delta x > 0$. Using the Normalized Gradient method to approximate Q on the grid $G_{\Delta x}$, we obtain the approximate solution

$$\widetilde{u}(t,x) = (\mathcal{M}_{P(t)}\widetilde{Q})(x),$$

where $\|Q - \widetilde{Q}\|_{L^2(G_{\Delta x})} \leq \eta$, and we extend $\widetilde{Q}$ by interpolation on the full box G, and by zero on $\mathbb{R}^2 \setminus G$. The gradient $\nabla\widetilde{Q}$ is assumed to be computed on $G_{\Delta x}$ with a pointwise error of order $\mathcal{O}(\Delta x^\sigma)$ for some $\sigma > 0$ (e.g. $\sigma = 1$ for finite differences of order 1).

Lemma 6 (Normalized Gradient method estimates). *Let* $\varepsilon_p(\alpha) := \|Q|y|^p\|^2_{L^2(\mathbb{R}^2\setminus G)}$ *and* $\epsilon(\alpha) := \|\nabla Q\|^2_{L^2(\mathbb{R}^2\setminus G)}$. *We have the following estimates:*

$$\|Q - \widetilde{Q}\|_{L^2(\mathbb{R}^2)} \lesssim \sqrt{\eta^2 + (\alpha\Delta x)^2 + \varepsilon_0(\alpha)},$$

$$\|(Q - \widetilde{Q})\nabla M\|_{L^2(\mathbb{R}^2)} \lesssim \sqrt{(1+|\alpha|^2)\left(\eta^2 + (\alpha\Delta x)^2\right) + \varepsilon_1(\alpha)},$$

$$\|M\nabla(Q - \widetilde{Q})\|_{L^2(\mathbb{R}^2)} \lesssim \sqrt{\alpha^2(\Delta x)^\sigma + C'(\alpha\Delta x)^2 + \epsilon(\alpha)},$$

where the implicit constants depend on the modulation parameters at time $t = 0$ *as well as on*

$$\|\nabla Q\|_{L^\infty}, \quad \|\nabla^2 Q\|_{L^\infty}, \quad \|\nabla\widetilde{Q}\|_{L^\infty}, \quad \text{and} \quad \|\nabla^2\widetilde{Q}\|_{L^\infty},$$

that are assumed to be bounded independently on Δx, α *and* η.

Proof. By a Taylor-Lagrange expansion, for $x \in [-\Delta x, \Delta x]^2$, there exist $\theta, \widetilde{\theta} \in (-\Delta x, \Delta x)$ such that

$$Q(x) - \widetilde{Q}(x) = Q(0) + x \cdot \nabla Q(\theta) - \left(\widetilde{Q}(0) + x \cdot \nabla\widetilde{Q}(\widetilde{\theta})\right).$$

We get

$$\begin{aligned}
&\int_{[-\Delta x,\Delta x]^2} \left|Q(x) - \widetilde{Q}(x)\right|^2 dx \\
&= \int_{[-\Delta x,\Delta x]^2} \left|Q(0) + x \cdot \nabla Q(\theta) - \left(\widetilde{Q}(0) + x \cdot \nabla\widetilde{Q}(\widetilde{\theta})\right)\right|^2 dx \\
&\leq 2\int_{[-\Delta x,\Delta x]^2} \left|Q(0) - \widetilde{Q}(0)\right|^2 dx + 2\int_{[-\Delta x,\Delta x]^2} \left|x \cdot \nabla Q(\theta) - x \cdot \nabla\widetilde{Q}(\widetilde{\theta})\right|^2 dx \\
&\leq 2\int_{[-\Delta x,\Delta x]^2} \left|Q(0) - \widetilde{Q}(0)\right|^2 dx + 2\int_{[-\Delta x,\Delta x]^2} |x|^2 \left|\nabla Q(\theta) - \nabla\widetilde{Q}(\widetilde{\theta})\right|^2 dx \\
&\leq 2(\Delta x)^2 \left|Q(0) - \widetilde{Q}(0)\right|^2 + 4C(\Delta x)^4,
\end{aligned}$$

where C depends on L^∞ on ∇Q and $\nabla\widetilde{Q}$. Then, similarly

$$\begin{aligned}\|Q-\widetilde{Q}\|^2_{L^2(G)} &= \int_G \left|Q-\widetilde{Q}\right|^2 dx\\ &= \sum_{g\in G_{\Delta x}} \int_{g_1}^{g_1+\Delta x}\int_{g_2}^{g_2+\Delta x} \left|Q(x)-\widetilde{Q}(x)\right|^2 dx\\ &\le 2(\Delta x)^2 \sum_{g\in G_{\Delta x}} \left|Q(g)-\widetilde{Q}(g)\right|^2 + 4C\sum_{g\in G}(\Delta x)^4\\ &\lesssim \|Q-\widetilde{Q}\|^2_{L^2(G_{\Delta x})} + C\alpha^2(\Delta x)^2 \lesssim \eta^2 + C\alpha^2(\Delta x)^2. \quad (3.17)\end{aligned}$$

We recall that $\widetilde{Q}$ is extended by zero on $\mathbb{R}^2\setminus G$, which shows the first estimate.

For the second estimate, we have

$$\begin{aligned}&\left\|(Q-\widetilde{Q})\nabla M\right\|^2_{L^2(\mathbb{R}^2)} = \left\|(Q-\widetilde{Q})(L\beta+By)M\right\|^2_{L^2(\mathbb{R}^2)}\\ &= \left\|(Q-\widetilde{Q})\left(\beta+\frac{B}{L}y\right)\right\|^2_{L^2(\mathbb{R}^2)}\\ &\le 2\left(|\beta|^2+2\left|\frac{B}{L}\alpha\right|^2\right)\left\|Q-\widetilde{Q}\right\|^2_{L^2(G)} + \left\|(Q-\widetilde{Q})\left(\beta+\frac{B}{L}y\right)\right\|^2_{L^2(\mathbb{R}^2\setminus G)}.\end{aligned}$$

By Lemma 3, there exists $C_1 = C(\beta(0), L(0), B(0))$ such that

$$\left\|(Q-\widetilde{Q})\nabla M\right\|^2_{L^2(\mathbb{R}^2)} \le 2C_1\left(1+|\alpha|^2\right)\left\|Q-\widetilde{Q}\right\|^2_{L^2(G)} + C_1\|Qy\|^2_{L^2(\mathbb{R}^2\setminus G)},$$

and we conclude using (3.17). The third estimate is obtained in a similar fashion:

$$\begin{aligned}\left\|M\nabla(Q-\widetilde{Q})\right\|^2_{L^2(\mathbb{R}^2)} &= \left\|\nabla_y(Q-\widetilde{Q})\right\|^2_{L^2(G)} + \epsilon(\alpha)\\ &\lesssim \left\|\nabla_y(Q-\widetilde{Q})\right\|^2_{L^2(G_{\Delta x})} + C'(\Delta x)^2\alpha^2 + \epsilon(\alpha),\end{aligned}$$

where C' depends on L^∞ bounds on the Hessian matrices of Q and $\widetilde{Q}$. Owing to the pointwise error estimate on the gradients we get

$$\left\|M\nabla(Q-\widetilde{Q})\right\|^2_{L^2(\mathbb{R}^2)} \lesssim \alpha^2(\Delta x)^\sigma + (\alpha\Delta x)^2 + \epsilon(\alpha),$$

which achieves the proof. □

Proposition 7. *The following estimates hold:*

$$\|u-\widetilde{u}\|_{L^2(\mathbb{R}^2)} \lesssim \sqrt{\eta^2+(\Delta x)^2\alpha+\varepsilon(\alpha)},$$

and

$$\begin{aligned}\|\nabla u-\nabla\widetilde{u}\|_{L^2(\mathbb{R}^2)} \lesssim &\sqrt{(1+|\alpha|^2)\left(\eta^2+(\alpha\Delta x)^2\right)+\varepsilon_1(\alpha)}\\ &+\sqrt{\alpha^2(\Delta x)^\sigma + C'(\alpha\Delta x)^2+\epsilon(\alpha)}.\end{aligned}$$

Proof. By a change of variables,

$$\|\widetilde{u}\|^2_{L^2(\mathbb{R}^2)} = \int_{\mathbb{R}^2} \left| \frac{1}{L} e^{i\gamma + iL\beta\cdot y(t,x) + \frac{iB}{2}|y(t,x)|^2} \widetilde{Q}(y(t,x)) \right|^2 dx = \int_{\mathbb{R}^2} \left|\widetilde{Q}(y)\right|^2 dy,$$

which is time-independent. For the L^2 norm estimate of $u - \widetilde{u}$,

$$\begin{aligned}\|u - \widetilde{u}\|^2_{L^2(\mathbb{R}^2)} &= \frac{1}{L^2}\|Q \circ y - \widetilde{Q} \circ y\|^2_{L^2(\mathbb{R}^d)} = \|Q - \widetilde{Q}\|^2_{L^2(\mathbb{R}^d)} \\ &\lesssim \eta^2 + (\alpha \Delta x)^2 + \varepsilon(\alpha).\end{aligned}$$

For the gradient of the difference,

$$\begin{aligned}&\|\nabla u - \nabla \widetilde{u}\|_{L^2(\mathbb{R}^2)} \\ &= \left\| Q \circ y \nabla_x M + M \nabla_x Q \circ y - \widetilde{Q} \circ y \nabla_x M - M \nabla_x \widetilde{Q} \circ y \right\|_{L^2(\mathbb{R}^2)} \\ &= \frac{1}{L}\left\| Q \circ y \nabla_y M + M \nabla_y Q \circ y - \widetilde{Q} \circ y \nabla_y M - M \nabla_y \widetilde{Q} \circ y \right\|_{L^2(\mathbb{R}^2)} \\ &= \left\| Q \nabla M + M \nabla Q - \widetilde{Q} \nabla M - M \nabla \widetilde{Q} \right\|_{L^2(\mathbb{R}^2)} \\ &\leq \left\| Q \nabla M - \widetilde{Q} \nabla M \right\|_{L^2(\mathbb{R}^2)} + \left\| M \nabla Q - M \nabla \widetilde{Q} \right\|_{L^2(\mathbb{R}^2)}.\end{aligned}$$

We get the claimed estimate by using Lemma 6. □

4. Numerical results

The efficiency of the proposed method is assessed based on the relative evolution of conservative quantities. For a quantity $q(t)$, its relative evolution is defined as

$$q_{\mathrm{rel}}(t) := \frac{q(t) - q(0)}{q(0)}, \quad t \geq 0.$$

The numerical simulations are all done over a time interval $[0, T]$, $T = 100$, discretized with a stepsize $\Delta t = 10^{-1}$.

For the soliton approximation, we have chosen a grid of $M = 100$ points in each direction, and looked for the soliton approximation over $[-5, 5] \times [-5, 5]$. The Normalized Gradient method is applied until two successive iterations $\psi^{(\ell)}$ and $\psi^{(\ell+1)}$ satisfy $\|\psi^{(\ell)} - \psi^{(\ell+1)}\|_{L^2} \leq \eta = 10^{-9}$.

For the computation of the conservative quantities, we approximate the first-order derivatives of the soliton using the Discrete Fourier method – in practice, using the FFT – and the derivatives of $M_{P(t)}(x)$ are computed exactly.

The results are not compared with those of a grid-based method, because it is not known beforehand what size should the grid be: even if

$X(0) = (0,0)$, if $\beta(0) = (a,a)$ for a large then the soliton will move far away from the origin. A purely grid-based method has no means of knowing what size the grid should be only based on the initial condition. The same holds for the stepsize of the grid: the grid has to account for the fast variations of the solutions, but it cannot know the fastest variations the solution will have at any time $t > 0$ only based on the initial condition.

4.1. *A single moving soliton*

The initial condition is made of a single modulated soliton, with parameters $X(0) = (5,0)$, $\beta(0) = (0,2)$, $L(0) = 1.5$, $B(0) = 0$, $\gamma(0) = 0$. The nonlinearity parameter is $\mu = 1$.

For this situation, we expect the algorithm to have excellent performances: the Dirac-Frenkel splitting step has right-hand side $S = 0$ so the approximate parameter derivatives will be zero, and thus the time evolution of the soliton is performed exactly. The numerical results are given in Figure 1.

However, we note that the conservative quantities are not preserved up to machine precision. This is due to the fact that in our implementation, the stepsize δx of the grid used to compute the conservative quantities is chosen to be constant and equal to the stepsize Δx of the grid used for the Normalized Gradient method. In the case of a single soliton, a better choice would be to use $\delta x = L\Delta x$, so that the computation of the conservative quantities would simply be a scaling of the initial condition. Numerical results – not reported here – showed that with this better choice of δx, the conservation is of order 10^{-12} for the L^2 norm and of order 10^{-10} for the energy and momentum. However this perfect choice of δx is not applicable in the general setting of several solitons, and in this case the most natural stepsize is $\delta x = \Delta x$. This explains why the conservative quantities in Figure 1 are not preserved down to machine precision.

4.2. *Rotating solitons*

The initial condition is made of two modulated solitons, one rotating clockwise close to the origin and one rotating anti-clockwise far from the origin.

The nonlinearity parameter is $\mu = 1$, and the initial modulations parameters are

$$\begin{array}{lllll} L^1 = 1 & B^1 = 0 & \gamma^1 = 0 & X^1 = (2,-2) & \beta^1 = (X^1)^\perp, \\ L^2 = 0.7 & B^2 = 0 & \gamma^2 = 0 & X^2 = (-4,4) & \beta^2 = (X^2)^\perp, \end{array}$$

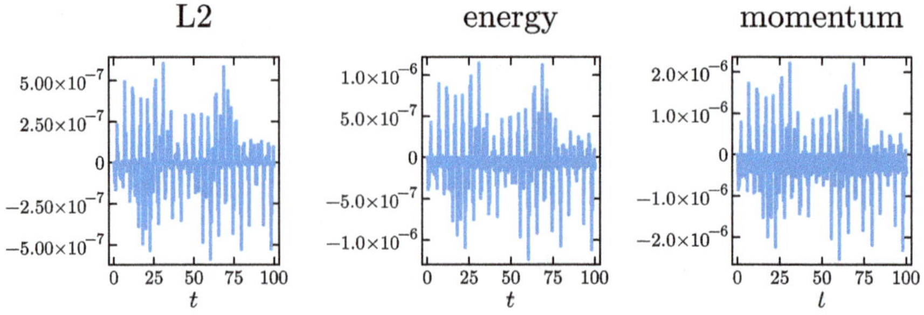

Fig. 1. Time evolution of the relative conservative quantities: L^2 norm (left), energy (center) and momentum (right), for the single modulated soliton. Nonlinearity parameter $\mu = 1$.

where $(x_1, x_2)^\perp = (-x_2, x_1)$, and $\mu = 1$. The parameters are chosen so that the two modulated solitons only interact weakly: weak interactions when $\mu = 1$ mean that the numerical grids associated to u^1 and u^2 have small overlaps. The results are given in Figure 2. Because of the weak interactions, the approximation due to the Dirac-Frenkel principle is small and we can have good conservation properties.

Also, a visual representation of the approximate solution at different times is given in Figure 3.

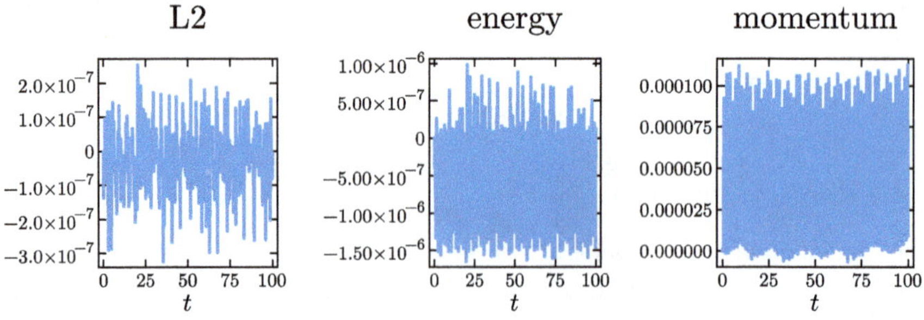

Fig. 2. Time evolution of the relative conservative quantities: L^2 norm (left), energy (center) and momentum (right), for the two modulated solitons with weak interactions. Nonlinearity parameter $\mu = 1$.

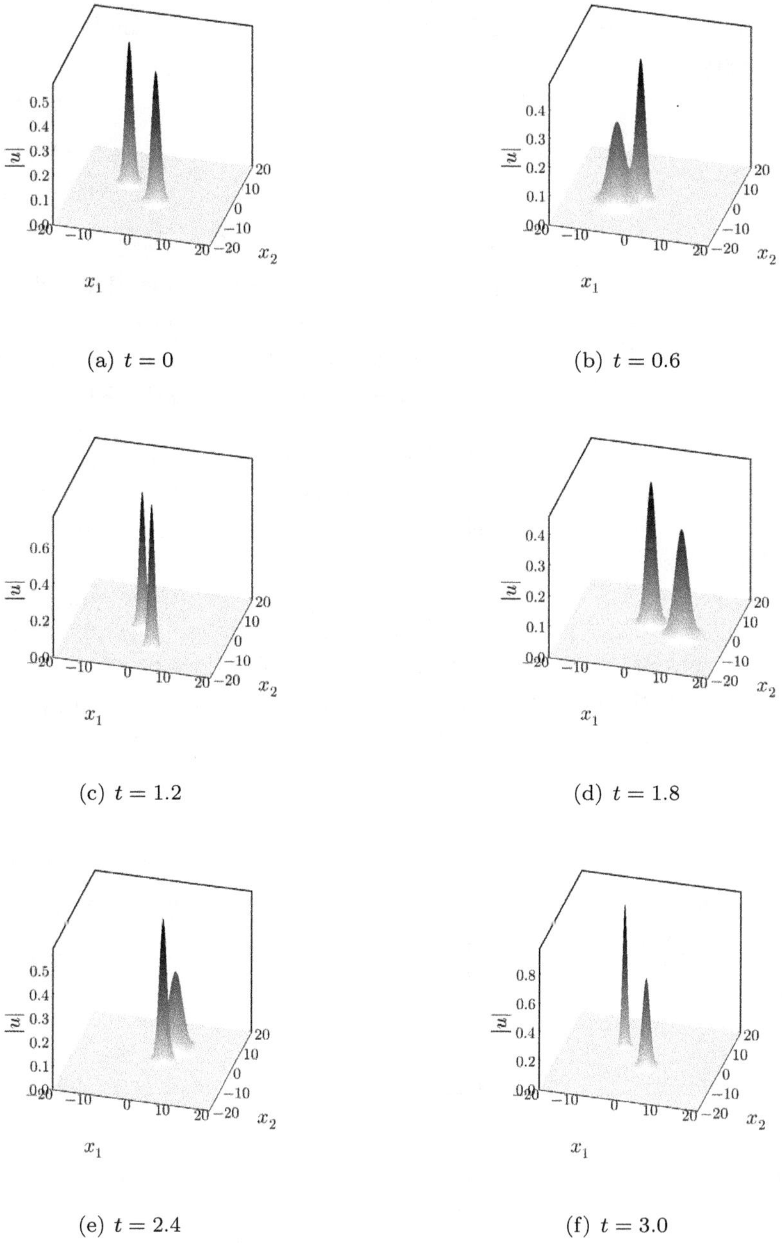

Fig. 3. Absolute value of the approximate solution to the weakly-interacting rotating soliton example.

4.3. *Bad scenario*

The third example consists in two solitons interacting heavily. The initial parameters are

$$\begin{array}{lllll} L^1 = 1 & B^1 = 0 & \gamma^1 = 0 & X^1 = (2, -2) & \beta^1 = (X^1)^\perp, \\ L^2 = 0.7 & B^2 = 0 & \gamma^2 = 0 & X^2 = (-4, 4) & \beta^2 = (X^2)^\perp, \end{array}$$

where $(x_1, x_2)^\perp = (-x_2, x_1)$, and they are chosen so that the grids associated to the two modulated solitons have significant overlaps.

We give in Figure 4 the results obtained with a nonlinearity parameter $\mu = 1$, and in Figure 5 the results obtained with a nonlinearity parameter $\mu = 10^{-3}$.

Because of the large overlaps of the two grids, the approximation error due to the Dirac-Frenkel principle – which imposes the ansatz (3.1) – is large when $\mu = 1$, and thus the numerical solution is a poor approximation of the true solution. The very rough approximations we make in this case can be seen on the relative time evolution of the monitored quantities, and in particular with the energy and momentum. When $\mu = 10^{-3}$, the nonlinearity is smaller and thus the true solution is closer to the ansatz (3.1), which means the approximation error due to the Dirac-Frenkel principle is smaller. These results underline that the method presented is a good choice in the case of weakly interacting solitons.

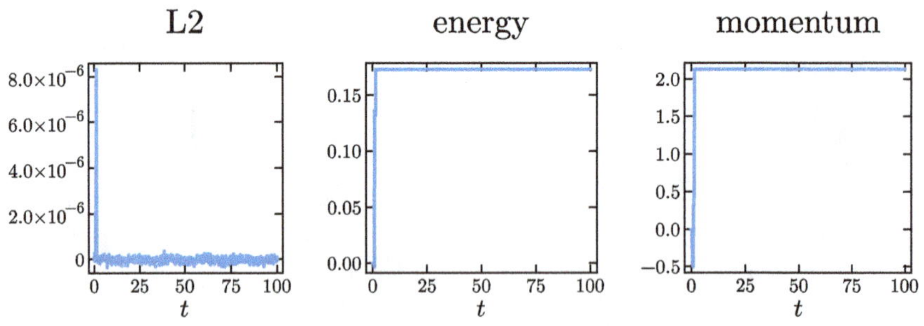

Fig. 4. Time evolution of the relative conservative quantities: L^2 norm (left), energy (center) and momentum (right), for the two modulated solitons with strong interactions. Nonlinearity parameter $\mu = 1$.

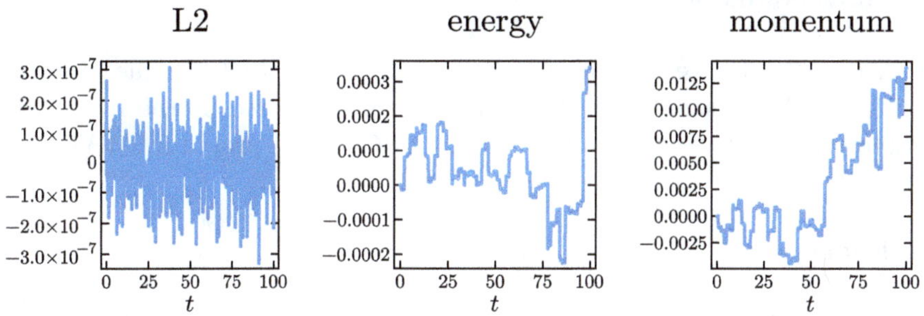

Fig. 5. Time evolution of the relative conservative quantities: L^2 norm (left), energy (center) and momentum (right), for the two modulated solitons with strong interactions. Nonlinearity parameter $\mu = 10^{-3}$.

5. Conclusion

The goal of this work was to use the theoretical idea of solution modulation for deriving a numerical algorithm.

When a single soliton is considered the modulation parameters are known exactly for all times, and the approximate numerical solution is very close to the true solution – up to some error that depends on the numerical parameters. Moreover the conservative quantities are very well-preserved for the approximate solution, and it is illustrated on a numerical example.

When more than one soliton is considered, we are interested in keeping the ansatz that describes the approximate solution as a sum of modulated solitons, using the Dirac-Frenkel principle. If the modulated solitons only have weak interactions – either they are far from each other or the nonlinearity parameter is small – the approximation error due to the Dirac-Frenkel principle is small and the conservative quantities of the approximate solution are still preserved up to a satisfying error. However, if the solitons interact heavily – either they are close to each other or the nonlinearity parameter is large – the error of the Dirac-Frenkel principle is large and a lot of information is lost during the projection step. Therefore, the conserved quantities of the approximate solution in this case are very poorly preserved.

By studying the error of the approximate soliton obtained via the Normalized Gradient method, we got an estimate on the error of the approximate solution. However, this is done only in the case of a single soliton, because when several solitons are interacting one has to study the error made by the Dirac-Frenkel principle.

Acknowledgments

E.F. and Y.L.H. were sponsored by the Centre Henri Lebesgue, program ANR-11-LABX-0020-01. E.F. and P.R. were sponsored by the Inria Associated team *Bubbles*. P.R supported by ERC/UKRI program SWAT.

References

[1] H. Li, Q. Zhai, and J. Z. Y. Chen, Neural-network-based multistate solver for a static Schrödinger equation, *Physical Review A*. **103**(3), 032405 (2021).

[2] C. Lasser and C. Lubich, Computing quantum dynamics in the semiclassical regime, *Acta Numerica*. **29**, 229–401 (2020).

[3] L. F. Mollenhauer and J. Gordon, P., *Solitons in Optical Fibers*. Elsevier (2006).

[4] G. Agrawal, *Nonlinear Fiber Optics*. Elsevier (2013).

[5] T. Cazenave, *Semilinear Schrödinger Equations*. vol. 10, *Courant Lecture Notes*, American Mathematical Society, Providence, Rhode Island (2003).

[6] R. Carles, *Semi-Classical Analysis for Nonlinear Schrödinger Equations: WKB Analysis, Focal Points, Coherent States*, 2nd edition edn. World Scientific, New Jersey (2021).

[7] Q. Chang, E. Jia, and W. Sun, Difference schemes for solving the generalized nonlinear Schrödinger equation, *Journal of Computational Physics*. **148**(2), 397–415 (1999).

[8] M. M. Cerimele, M. L. Chiofalo, F. Pistella, S. Succi, and M. P. Tosi, Numerical solution of the Gross-Pitaevskii equation using an explicit finite-difference scheme: An application to trapped Bose-Einstein condensates, *Physical Review E*. **62**(1), 1382–1389 (2000).

[9] W. Bao, D. Jaksch, and P. A. Markowich, Numerical solution of the Gross–Pitaevskii equation for Bose–Einstein condensation, *Journal of Computational Physics*. **187**(1), 318–342 (2003).

[10] E. Faou, *Geometric Numerical Integration and Schrödinger Equations*. Zurich Lectures in Advanced Mathematics, European Mathematical Society, Zürich, Switzerland (2012).

[11] W. Bao, H. Li, and J. Shen, A generalized-Laguerre–Fourier–Hermite pseudospectral method for computing the dynamics of rotating Bose-Einstein condensates, *SIAM Journal on Scientific Computing*. **31**(5), 3685–3711 (2009).

[12] X. Antoine, W. Bao, and C. Besse, Computational methods for the dynamics of the nonlinear Schrödinger/Gross-Pitaevskii equations, *Computer Physics Communications*. **184**(12), 2621–2633 (2013).

[13] J. Bernier, N. Crouseilles, and Y. Li, Exact splitting methods for kinetic and Schrödinger equations, *Journal of Scientific Computing*. **86**(1), 10 (2021).

[14] E. J. Heller, Time dependent variational approach to semiclassical dynamics, *The Journal of Chemical Physics*. **64**(1), 63–73 (1976).

[15] D. Huber and E. J. Heller, Generalized Gaussian wave packet dynamics, *The Journal of Chemical Physics.* **87**(9), 5302–5311 (1987).

[16] C. Lubich, *From Quantum to Classical Molecular Dynamics: Reduced Models and Numerical Analysis*, 1 edn. EMS Press (2008).

[17] K. Rowan, L. Schatzki, T. Zaklama, Y. Suzuki, K. Watanabe, and K. Varga, Simulation of a hydrogen atom in a laser field using the time-dependent variational principle, *Physical Review E.* **101**(2), 023313 (2020).

[18] T. Tsednee, B. Tsednee, and T. Khinayat. Numerical solution to the time-dependent Gross-Pitaevskii equation (2022).

[19] A. Klein, D. Jaksch, Y. Zhang, and W. Bao, Dynamics of vortices in weakly interacting Bose-Einstein condensates, *Physical Review A.* **76**(4), 043602 (2007).

[20] Y. Martel and P. Raphaël, Strongly interacting blow up bubbles for the mass critical nonlinear Schrödinger equation, *Annales scientifiques de l'École normale supérieure.* **51**(3), 701–737 (2018).

[21] E. Faou and P. Raphaël, On weakly turbulent solutions to the perturbed linear harmonic oscillator, *American Journal of Mathematics.* **145**(5), 1465–1507 (2023).

[22] R. D. Coalson and M. Karplus, Multidimensional variational Gaussian wave packet dynamics with application to photodissociation spectroscopy, *The Journal of Chemical Physics.* **93**(6), 3919–3930 (1990).

[23] G. A. Worth, M. A. Robb, and I. Burghardt, A novel algorithm for non-adiabatic direct dynamics using variational Gaussian wavepackets, *Faraday Discussions.* **127**, 307 (2004).

[24] L. Adamowicz, S. Kvaal, C. Lasser, and T. B. Pedersen, Laser-induced dynamic alignment of the HD molecule without the Born–Oppenheimer approximation, *The Journal of Chemical Physics.* **157**(14), 144302 (2022).

[25] O. Kavian and F. B. Weissler, Self-similar solutions of the pseudo-conformally invariant nonlinear Schrödinger equation., *Michigan Mathematical Journal.* **41**(1) (1994).

[26] P. J. Olver, *Applications of Lie Groups to Differential Equations.* vol. 107, *Graduate Texts in Mathematics*, Springer New York, New York, NY (1986).

[27] F. Merle and P. Raphaël, The blow-up dynamic and upper bound on the blow-up rate for critical nonlinear Schrödinger equation, *Annals of Mathematics.* **161**(1), 157–222 (2005).

[28] F. Merle, P. Raphaël, I. Rodnianski, and J. Szeftel. On the implosion of a three dimensional compressible fluid (2020).

[29] P. Gérard, E. Lenzmann, O. Pocovnicu, and P. Raphaël, A two-soliton with transient turbulent regime for the cubic half-wave equation on the real line, *Annals of PDE.* **4**(1), 7 (2018).

[30] B. Leimkuhler and S. Reich, *Simulating Hamiltonian Dynamics*, 1 edn. Cambridge University Press (2005).

[31] E. Hairer, C. Lubich, and G. Wanner, *Geometric Numerical Integration: Structure-Preserving Algorithms for Ordinary Differential Equations*, 2nd ed edn. Number 31 in Springer Series in Computational Mathematics, Springer, Berlin ; New York (2006).

[32] E. Faou and T. Jézéquel, Convergence of a normalized gradient algorithm for computing ground states, *IMA Journal of Numerical Analysis.* **38**(1), 360–376 (2018).

[33] G. Hämmerlin and K.-H. Hoffman, *Numerical Mathematics.* Undergraduate Texts in Mathematics, Readings in Mathematics, Springer New York, New York, NY (1991).

[34] C. De Boor, *A Practical Guide to Splines: With 32 Figures*, rev. ed edn. Number v. 27 in Applied Mathematical Sciences, Springer, New York (2001).

[35] E. Sonnendrücker, J. Roche, P. Bertrand, and A. Ghizzo, The semi-Lagrangian method for the numerical resolution of the Vlasov equation, *Journal of Computational Physics.* **149**(2), 201–220 (1999).

https://doi.org/10.1142/9789819816620_0006

Chapter 6

Exponential-Type Fourier Integrator for KdV Equations with Rough Data: A Review

Ying He*, Yifei Wu† and Xiaofei Zhao‡

**Beijing Institute of Mathematical Sciences and Applications, 101408 Beijing, China, and Yau Mathematical Sciences Center, Tsinghua University, 100084, Beijing, China*
heymath@bimsa.cn

†School of Mathematical Sciences Nanjing Normal University, 210023, Nanjing, China
yerfmath@gmail.com

‡School of Mathematics and Statistics & Computational Sciences Hubei Key Laboratory, Wuhan University, 430072, Wuhan, China
matzhxf@whu.edu.cn

In this chapter, we review the exponential-type integrators proposed in recent years based on frequency analysis for solving the KdV-type equations under rough data. The required regularity of the class of methods for accuracy is lower than classical approaches. This covers works for classical KdV, mKdV and gKdV equations.

Contents

1. Introduction

In this chapter, we consider the low-regularity integration approach to solve numerically the KdV-type equations under rough initial data on a torus:

$$\begin{cases} \partial_t u(t,x) + \partial_x^3 u(t,x) + \frac{1}{k}\partial_x\big(u(t,x)^k\big) = 0, & x \in \mathbb{T},\ t > 0, \\ u(0,x) = u_0(x), & x \in \mathbb{T}, \end{cases} \tag{1.1}$$

where $\mathbb{T} = (0, 2\pi)$, $u(t,x) : \mathbb{R}^+ \times \mathbb{T} \to \mathbb{R}$ is the unknown and $u_0 \in H^s(\mathbb{T})$ is the given initial data with some $0 \leq s < \infty$. Here the power index $k > 0$ is an arbitrary given integer and based on its value, (1.1) is usually classified as: the (classical) KdV equation and the modified KdV equation (mKdV) for $k = 2$ and $k = 3$, respectively; the generalized KdV equation (gKdV) for all $k \geq 4$. As a classical mathematical model to describe the dynamics of waves on shallow water surface, equation (1.1) has drawn extensive research interests from many different aspects.

Theoretically, the well-posedness of the Cauchy problem (1.1) has been intensively investigated, particularly for initial data $u_0 \in H^s$ with different regularity. Let us give a brief review here. The global well-posedness theory for the classical KdV and the mKdV have been established, respectively, for $s \geq -1$ [1] and $s \geq \frac{1}{2}$ [2]. For the gKdV equation, the local well-posedness has been shown for $s \geq \frac{1}{2}$ firstly by Colliander, Keel, Staffilani, Takaoka and Tao [3], and Bao and Wu [4] later obtained the global well-posedness in the quartic nonlinearity case of (1.1). By assuming that the solution is smooth enough, different kinds of discretization techniques including finite difference methods [5, 6], operator splittings [7–10], finite elements [11, 12], spectral methods [13–15] and discontinuous Galerkin methods [16–18] have been proposed and analyzed to solve the KdV-type model (1.1). However, when the solutions of the model are not smooth enough, e.g., the rough data case, the above numerical methods will lose their accuracy and then become less effective. In practice, the rough initial data may be caused by multiple reasons such as measurements (measurement data produced by, for example, inaccurate instruments or incorrect readings) or noise [19–21].

To tackle the rough data case and increase the temporal accuracy, a so-called low-regularity integrator was introduced by Ostermann and Schratz for nonlinear Schrödinger equations [22]. Compared with the traditional numerical discretizations, the low-regularity integrator is favored for reaching the optimal convergence for less regular solutions. Ever since then, a trend has been started to design low-regularity integrators for important dispersive models [23–33], where the models possess no internal mechanism to smoothen the rough data. As one of the big three dispersive models, the KdV-type equations (1.1) have received special attentions. Hofmanová and Schratz [34] considered an exponential-type integrator [35], which is based on the twisted variable $\mathfrak{u} = e^{\partial_x^3 t}u$ and the Duhamel formula of (1.1) at $t_n = n\tau$ with $\tau > 0$ the time step:

$$\mathfrak{u}(t_n+\tau,x) = \mathfrak{u}(t_n,x) - \frac{1}{k}\int_0^\tau e^{(t_n+s)\partial_x^3}\partial_x\left(e^{-(t_n+s)\partial_x^3}\mathfrak{u}(t_n+s,x)\right)^k ds.$$

By letting $\mathfrak{u}(t_n+s,x) \approx \mathfrak{u}(t_n,x)$ in the above, the key ingredient is to find the integration for s which in Fourier frequency space reads

$$\frac{1}{k}\int_0^\tau \sum_{l=l_1+\cdots+l_k} e^{-i(t_n+s)(l^3-l_1^3-l_2^3-\cdots-l_k^3)}(il)\widehat{\mathfrak{u}}(t_n,l_1)\widehat{\mathfrak{u}}(t_n,l_2)\cdots\widehat{\mathfrak{u}}(t_n,l_k)ds.$$

When $k=2$, owing to a key factorization formula $l^3 - l_1^3 - l_2^3 = 3ll_1l_2$ for the phase function, the above integral can not only be evaluated exactly but can also be defined point-wisely in the physical space. As proved in [34], the resulting integrator is first-order accurate in H^1-norm for solving the classical KdV equation with initial data from H^3. The scheme therefore requires two additional bounded derivatives of the solution to reach the first order accuracy, which is indeed lower than the requirement of the splitting method [7] or the finite difference method [5]. This requirement is to some extend essential for direct integration methods due to the Burgers-type nonlinearity in the KdV equation [36]. The study for the second order scheme has been done in [37]. However, the technique from [34] can not work for $k > 2$. Extra developments have then been done by Wu, Zhao et al. for general KdV equations in recent years.

In this chapter, we shall review firstly the work for gKdV equations [24], which covers all $k \geq 2$. The exponential-type integrator therein was proposed based on a gauge-transform, and it can achieve the first order accuracy with two additional bounded derivatives. Then, we will review the cases for classical KdV and mKdV equations, where the regularity requirement can be further reduced. For the classical KdV equation, a first scheme that costs only one derivative was proposed [38]. For the mKdV equation case, the celebrated Muira transform helps to derive the corresponding low-regularity integrator [39]. These results will be reviewed in subsequent sections.

The rest of the chapter is organized as follows. In Section 2, we will give the notations commonly used in the presentations. In Section 3, we review the results for gKdV equations. The results of KdV and mKdV equations will be reviewed in Section 4 and Section 5, respectively.

2. Preliminaries

We give some notations and tool lemmas that will be used later for the derivation and analysis of the scheme.

2.1. *Notation*

To discretize the time axis, we denote by $\tau = \Delta t > 0$ the time step and $t_n = n\tau$ for $n \in \mathbb{N}$ as the grid points. For two quantities A and B, $A \lesssim B$ denotes $A \leq CB$ for some constant $C > 0$ whose value may vary line by line but is independent of τ, n.

The Fourier transform of a function $f(x)$ on $\mathbb{T}$ is defined by

$$\mathcal{F}(f) = \widehat{f}(l) = \frac{1}{2\pi}\int_{\mathbb{T}} \mathrm{e}^{-ilx} f(x)\, dx, \text{ with the inversion } \quad f(x) = \sum_{l\in\mathbb{Z}} \mathrm{e}^{ilx}\widehat{f}(l).$$

Note that $\widehat{f}(0) = \frac{1}{2\pi}\int_{\mathbb{T}} f\,dx$, gives the average. The following classical properties hold:

$$\|f\|_{L^2(\mathbb{T})} = \sqrt{2\pi}\Big(\sum_{l\in\mathbb{Z}} |\widehat{f}(l)|^2\Big)^{\frac{1}{2}} \text{ (Plancherel)};$$

$$\langle f, g\rangle = \int_{\mathbb{T}} f(x)\overline{g(x)}\,dx = 2\pi\sum_{l\in\mathbb{Z}} \widehat{f}(l)\overline{\widehat{g}(l)} \text{ (Parseval)};$$

$$\widehat{(fg)}(l) = \sum_{l_1\in\mathbb{Z}} \widehat{f}(l-l_1)\widehat{g}(l_1) \text{ (Convolution)}.$$

The Sobolev space $H^s(\mathbb{T})$ for $s \geq 0$ has the equivalent norm:

$$\|f\|_{H^s(\mathbb{T})} = \|J^s f\|_{L^2(\mathbb{T})} = \Big(\sum_{l\in\mathbb{Z}} (1+l^2)^s|\widehat{f}(l)|^2\Big)^{\frac{1}{2}},$$

where we denote the operator

$$J^s = (1-\partial_{xx})^{\frac{s}{2}}.$$

Denote $\langle\cdot\rangle = (1+|\cdot|^2)^{\frac{1}{2}}$ and $|\nabla|^{s_0} = (-\partial_{xx})^{\frac{s_0}{2}}$ for $s_0 > 0$. For short, we denote the functional space

$$L_t^\infty H_x^{s_0} = L^\infty((0,T);H^{s_0}(\mathbb{T}))$$

for any $s_0 \geq 0$. Here $T > 0$ is the final time of the computation. Moreover, we denote by ∂_x^{-1} the inversion operator

$$\widehat{\partial_x^{-1} f}(\xi) = \begin{cases} (il)^{-1}\widehat{f}(l), & \text{when } l \neq 0, \\ 0, & \text{when } l = 0. \end{cases}$$

For a space-time function $f(x,t)$, when there is no ambiguity we will omit the spatial variable x and denote $f(t) = f(x,t)$ for simplicity.

3. Gauge-transformed exponential integrators for gKdV equation

3.1. *Scheme*

In this section, we shall begin with the construction of our new exponential integrator for the gKdV equation (1.1) based on the gauge transform.

3.1.1. *Gauge transformation*

We define the so-called *gauge transform* for the solution of (1.1) as

$$w(t,x) := u\Big(t, x + \frac{1}{2\pi}\int_0^t \int_{\mathbb{T}} u(\rho,x)^{k-1} dx d\rho\Big), \quad x \in \mathbb{T},\ t \geq 0. \tag{3.1}$$

Note that $u(t,x)$ is a periodic function of x on $\mathbb{T}$ for every t, so $\int_{\mathbb{T}} (u(t,x))^{k-1}\, dx = \int_{\mathbb{T}} (w(t,x))^{k-1}\, dx$. The transformation is invertible. Indeed, we have the inverse transform as

$$u(t,x) = w\Big(t, x - \frac{1}{2\pi}\int_0^t \int_{\mathbb{T}} w(\rho,x)^{k-1} dx d\rho\Big). \tag{3.2}$$

Under the gauge transform (3.1), equation (1.1) is transformed into the following equivalent form:

$$\partial_t w(t,x) + \partial_x^3 w(t,x) = -\partial_x w(t,x) w(t,x)^{k-1} + \partial_x w(t,x) \frac{1}{2\pi}\int_{\mathbb{T}} w(t,x)^{k-1} dx, \tag{3.3}$$

where $x \in \mathbb{T},\ t > 0$ with the initial value $w(0,x) = u_0(x)$ unchanged.

To the best of our knowledge, the transform of the form (3.1) was firstly considered in the work of Staffilani [40] for the well-posedness analysis. Here, we find it as a vital tool to eliminate the resonance term coming up from the numerical approximation of the nonlinearity later, which is the key to avoiding the stability issue. This will become clear in the detailed derivation below and the proof in the next section. Our strategy is to first construct the numerical solution to (3.3), and then obtain the one of (1.1) by the inverse gauge transformation.

3.1.2. *Derivation of the scheme*

Based on (3.3), we construct the numerical solution in the framework of the exponential-type time integrator. The Duhamel formula of (3.3) at some t_n reads

$$w(t_{n+1}) = \mathrm{e}^{-\tau\partial_x^3} w(t_n) - \int_0^\tau \mathrm{e}^{-(\tau-s)\partial_x^3}\left[\partial_x w(t_n+s)\cdot w(t_n+s)^{k-1}\right] ds$$
$$+ \int_0^\tau \mathrm{e}^{-(\tau-s)\partial_x^3}\left[\partial_x w(t_n+s)\cdot \frac{1}{2\pi}\int_{\mathbb{T}} w(t_n+s)^{k-1}\,dx\right] ds.$$

Now by introducing the new twisted variable $v(t) := \mathrm{e}^{t\partial_x^3} w(t)$, the above formula can be rewritten as

$$\begin{aligned} &v(t_{n+1}) \\ &= v(t_n) + \int_0^\tau \partial_x v(t_n+s)\,\frac{1}{2\pi}\int_{\mathbb{T}}\left(\mathrm{e}^{-(t_n+s)\partial_x^3} v(t_n+s)\right)^{k-1} dx ds \\ &\quad - \int_0^\tau \mathrm{e}^{(t_n+s)\partial_x^3}\left[\left(\mathrm{e}^{-(t_n+s)\partial_x^3}\partial_x v(t_n+s)\right)\left(\mathrm{e}^{-(t_n+s)\partial_x^3} v(t_n+s)\right)^{k-1}\right] ds. \end{aligned} \tag{3.4}$$

Applying the approximation $v(t_n+s) \approx v(t_n)$, we have

$$v(t_{n+1}) = v(t_n) + I_1(t_n) + I_2(t_n) + R_1^n, \tag{3.5}$$

where we denote

$$I_1(t_n) := -\int_0^\tau \mathrm{e}^{(t_n+s)\partial_x^3}\left[\left(\mathrm{e}^{-(t_n+s)\partial_x^3}\partial_x v(t_n)\right)\cdot\left(\mathrm{e}^{-(t_n+s)\partial_x^3} v(t_n)\right)^{k-1}\right] ds, \tag{3.6a}$$

$$I_2(t_n) := \int_0^\tau \partial_x v(t_n)\cdot\frac{1}{2\pi}\int_{\mathbb{T}}\left(\mathrm{e}^{-(t_n+s)\partial_x^3} v(t_n)\right)^{k-1} dx ds, \tag{3.6b}$$

and

$$
\begin{aligned}
R_1^n := &- \int_0^\tau e^{(t_n+s)\partial_x^3} \Big[\Big(e^{-(t_n+s)\partial_x^3} \partial_x v(t_n+s) \Big) \cdot \Big(e^{-(t_n+s)\partial_x^3} v\,(t_n+s) \Big)^{k-1} \\
&\qquad - \Big(e^{-(t_n+s)\partial_x^3} \partial_x v(t_n) \Big) \cdot \Big(e^{-(t_n+s)\partial_x^3} v\,(t_n) \Big)^{k-1} \Big] ds \\
&+ \int_0^\tau \Big[\partial_x v\,(t_n+s) \cdot \frac{1}{2\pi} \int_{\mathbb{T}} \Big(e^{-(t_n+s)\partial_x^3} v\,(t_n+s) \Big)^{k-1} dx \\
&\qquad - \partial_x v\,(t_n) \cdot \frac{1}{2\pi} \int_{\mathbb{T}} \Big(e^{-(t_n+s)\partial_x^3} v\,(t_n) \Big)^{k-1} dx \Big] ds. \qquad (3.7)
\end{aligned}
$$

The term R_1^n would be the truncation term. We shall show later in the detailed analysis that $\|R_1^n\|_{H^\gamma} = \mathcal{O}(\tau^2)$ and it depends on the second order derivative of the solution. Then to derive the scheme based on (3.5), it remains to evaluate $I_1(t_n)$ and $I_2(t_n)$. In order to get a practically efficient scheme as we discussed before, one should avoid convolutions in the Fourier frequency space. A closed form with point-wise definition in the physical space would be the ultimate goal and the key for approximating $I_1(t_n)$ and $I_2(t_n)$, because that would possibly allow us to apply fast algorithms. This will be further cleared in the derivation below.

Before we step into the detailed calculations, let us first briefly describe the main ideas and their motivations. By taking the Fourier transform on $I_1(t_n)$, we get

$$
\begin{aligned}
\widehat{I_1}(t_n, l) = - \int_0^\tau \sum_{l=l_1+l_2+\cdots+l_k} & e^{-i(t_n+s)(l^3-l_1^3-l_2^3-\cdots-l_k^3)} (il_1) \\
& \widehat{v}(t_n, l_1)\widehat{v}(t_n, l_2) \cdots \widehat{v}(t_n, l_k) ds, \qquad (3.8)
\end{aligned}
$$

where $l, l_1, l_2, \ldots, l_k \in \mathbb{Z}$. As mentioned, the phase function $l^3 - l_1^3 - l_2^3 - \cdots - l_k^3$ is the obstacle if we integrate directly, since it cannot be factorized. To overcome this difficulty and meanwhile reduce the cost of derivatives, we need to find a new technique to approximate the exponential term

$$
e^{-is(l^3-l_1^3-l_2^3-\cdots-l_k^3)}.
$$

Here we aim for a first-order scheme with the cost of two derivatives at most. We begin by rewriting

$$
l^3 - l_1^3 - l_2^3 - \cdots - l_k^3 = \alpha + \beta,
$$

where

$$
\alpha = l^3 - l_1^3 - (l_2 + \cdots + l_k)^3, \quad \beta = (l_2 + \cdots + l_k)^3 - l_2^3 - \cdots - l_k^3. \quad (3.9)
$$

Then, we split the exponential term into (similar kind of splitting was introduced in [26]):

$$e^{-is(\alpha+\beta)} = e^{-is\alpha} + (e^{-is\beta} - 1)e^{-is\alpha},$$

and so (3.8) is split into

$$\begin{aligned}
&\widehat{I_1}(t_n, l)\\
&= -\sum_{l=l_1+l_2+\cdots+l_k} e^{-it_n(l^3-l_1^3-l_2^3-\cdots-l_k^3)}(il_1)\\
&\qquad \widehat{v}(t_n, l_1)\widehat{v}(t_n, l_2)\cdots\widehat{v}(t_n, l_k)\int_0^\tau e^{-is\alpha}ds \qquad (3.10a)\\
&\quad -\sum_{l=l_1+l_2+\cdots+l_k} e^{-it_n(l^3-l_1^3-l_2^3-\cdots-l_k^3)}(il_1)\\
&\qquad \widehat{v}(t_n, l_1)\widehat{v}(t_n, l_2)\cdots\widehat{v}(t_n, l_k)\int_0^\tau (e^{-is\beta} - 1)e^{-is\alpha}ds. \qquad (3.10b)
\end{aligned}$$

Noting $l = l_1 + l_2 + \cdots + l_k$, (3.10a) is factorable:

$$\alpha = 3ll_1(l_2 + \cdots + l_k).$$

If $l \neq 0$, $l_1 \neq 0$, and $l_2 + \cdots + l_k \neq 0$, we then find

$$\int_0^\tau e^{-is\alpha}ds = \frac{e^{-i\tau\alpha} - 1}{-i\alpha} = \frac{e^{-i\tau\alpha} - 1}{-i3ll_1(l_2 + \cdots + l_k)}, \qquad (3.11)$$

which implies that (3.10a) can be exactly defined pointwisely in the physical space. The three special cases $l = 0$ or $l_1 = 0$ or $l_2 + \cdots + l_k = 0$ will have to be considered separately for (3.8). Let us here quickly check the case $l_2 + \cdots + l_k = 0$ in (3.8), where we can find

$$\begin{aligned}
&\int_0^\tau \sum_{\substack{l=l_1+\ldots+l_k\\ l_2+\cdots+l_k=0}} e^{-i(t_n+s)(l^3-l_1^3-l_2^3-\cdots-l_k^3)}(il_1)\widehat{v}(t_n, l_1)\widehat{v}(t_n, l_2)\cdots\widehat{v}(t_n, l_k)ds\\
&= \sum_{\substack{l=l_1\\ l_2+\cdots+l_k=0}} (il_1)\widehat{v}(t_n, l_1)\int_0^\tau e^{-i(t_n+s)(-l_2^3-\cdots-l_k^3)}\widehat{v}(t_n, l_2)\cdots\widehat{v}(t_n, l_k)ds\\
&= \partial_x v(t_n)\int_0^\tau \frac{1}{2\pi}\int_{\mathbb{T}}\left(e^{-(t_n+s)\partial_x^3}v(t_n)\right)^{k-1}dxds.
\end{aligned}$$

The above term matches precisely with the $I_2(t_n)$ that we introduced in (3.6b), and so it will be canceled in (3.5) by $I_2(t_n)$. Note that the term $I_2(t_n)$ comes from (3.3) which is in fact induced by the gauge transformation. Now our motivation to introduce the transformation becomes clear.

If one works directly on (1.1) without using the gauge transformation, the term $\partial_x v(t_n)$ in the above will be explicitly present in the final scheme, which will cause unbalanced norms in the numerical solution from n to $n+1$ and consequently lead to instability. As for (3.10b), it can be treated directly as a truncation, since

$$\left|(\mathrm{e}^{-is\beta}-1)\mathrm{e}^{-is\alpha}\right| \leq s|\beta| \quad \text{and} \quad |\beta| \lesssim \sum_{j\neq k\neq 1} |l_j|^2|l_k|.$$

The truncation error is therefore $\mathcal{O}(\tau^2)$ with the cost of only two spatial derivatives.

Now we are ready to present the derivation for the scheme. By the above discussion, we divide (3.8) into the following terms:

$$\begin{aligned}
&\widehat{I}_1(t_n,l)\\
&=-\int_0^\tau \sum_{l=l_1+l_2+\cdots+l_k=0} \mathrm{e}^{-i(t_n+s)(\alpha+\beta)}(il_1)\widehat{v}(t_n,l_1)\widehat{v}(t_n,l_2)\cdots\widehat{v}(t_n,l_k)ds \qquad (3.12a)\\
&\quad-\int_0^\tau \sum_{\substack{l=l_1+l_2+\cdots+l_k\neq 0\\ l_1=0}} \mathrm{e}^{-i(t_n+s)(\alpha+\beta)}(il_1)\widehat{v}(t_n,l_1)\widehat{v}(t_n,l_2)\cdots\widehat{v}(t_n,l_k)ds \qquad (3.12b)\\
&\quad-\int_0^\tau \sum_{l=l_1} \mathrm{e}^{-i(t_n+s)(\alpha+\beta)}(il_1)\widehat{v}(t_n,l_1)\widehat{v}(t_n,l_2)\cdots\widehat{v}(t_n,l_k)ds \qquad (3.12c)\\
&\quad-\int_0^\tau \sum_{\substack{l=l_1+l_2+\cdots+l_k\neq 0\\ l_1\neq 0,\, l_2+\cdots+l_k\neq 0}} \mathrm{e}^{-i(t_n+s)(\alpha+\beta)}(il_1)\widehat{v}(t_n,l_1)\widehat{v}(t_n,l_2)\cdots\widehat{v}(t_n,l_k)ds. \qquad (3.12d)
\end{aligned}$$

For (3.12a), by the symmetry within $l_1,\ldots,l_k$, we find

$$\begin{aligned}
&(3.12a)\\
&=-\frac{1}{k}\int_0^\tau \sum_{l=l_1+l_2+\cdots+l_k=0} \mathrm{e}^{-i(t_n+s)(\alpha+\beta)}(il)\widehat{v}(t_n,l_1)\widehat{v}(t_n,l_2)\cdots\widehat{v}(t_n,l_k)ds\\
&=0.
\end{aligned}$$

It can be directly checked that (3.12b) $=0$ and (3.12c) $=-\widehat{I}_2(t_n,l)$. Hence, we have

$$\widehat{I}_1(t_n,l)+\widehat{I}_2(t_n,l)=(3.12\text{d}).$$

For (3.12d), it can be approximated as

$$
\begin{aligned}
&(3.12\text{d})\\
&= -\int_0^\tau \sum_{\substack{l=l_1+l_2+\cdots+l_k\neq 0\\ l_1\neq 0, l_2+\cdots+l_k\neq 0}} \mathrm{e}^{-it_n(\alpha+\beta)}\mathrm{e}^{-is\alpha}(il_1)\widehat{v}(t_n,l_1)\widehat{v}(t_n,l_2)\cdots\widehat{v}(t_n,l_k)ds\\
&\quad + \widehat{R_2^n}(l),
\end{aligned}
\tag{3.13}
$$

where

$$
\begin{aligned}
\widehat{R_2^n}(l) := -\int_0^\tau \sum_{\substack{l=l_1+l_2+\cdots+l_k\neq 0\\ l_1\neq 0, l_2+\cdots+l_k\neq 0}} &\mathrm{e}^{-it_n(\alpha+\beta)}\big(\mathrm{e}^{-is\beta}-1\big)\mathrm{e}^{-is\alpha}(il_1)\\
&\widehat{v}(t_n,l_1)\widehat{v}(t_n,l_2)\cdots\widehat{v}(t_n,l_k)ds.
\end{aligned}
\tag{3.14}
$$

Therefore, (3.13) combined with (3.11) yields

$$
\begin{aligned}
&(3.12\text{d})\\
&= \sum_{\substack{l=l_1+l_2+\cdots+l_k\neq 0\\ l_1\neq 0, l_2+\cdots+l_k\neq 0}} \mathrm{e}^{-it_n(\alpha+\beta)}\frac{\mathrm{e}^{-i\tau\alpha}-1}{3l(l_2+\cdots+l_k)}\widehat{v}(t_n,l_1)\widehat{v}(t_n,l_2)\cdots\widehat{v}(t_n,l_k)\\
&\quad + \widehat{R_2^n}(l).
\end{aligned}
$$

The restriction $l_1 \neq 0$ can be removed, since when $l_1 = 0$, we have $\alpha = 0$ in (3.9) and the corresponding term vanishes. Hence,

$$
\begin{aligned}
&(3.12\text{d})\\
&= \sum_{\substack{l=l_1+l_2+\cdots+l_k\neq 0\\ l_2+\cdots+l_k\neq 0}} \mathrm{e}^{-it_n(\alpha+\beta)}\frac{\mathrm{e}^{-i\tau\alpha}-1}{3l(l_2+\cdots+l_k)}\widehat{v}(t_n,l_1)\widehat{v}(t_n,l_2)\cdots\widehat{v}(t_n,l_k)\\
&\quad + \widehat{R_2^n}(l).
\end{aligned}
$$

The inverse Fourier transform of the above gives

$$
\begin{aligned}
&I_1(t_n)+I_2(t_n)\\
&= -\frac{1}{3}\mathrm{e}^{t_{n+1}\partial_x^3}\partial_x^{-1}\Big[\Big(\mathrm{e}^{-t_{n+1}\partial_x^3}v(t_n)\Big)\cdot\Big(\mathrm{e}^{-\tau\partial_x^3}\partial_x^{-1}\big(\mathrm{e}^{-t_n\partial_x^3}v(t_n)\big)^{k-1}\Big)\Big]\\
&\quad + \frac{1}{3}\mathrm{e}^{t_n\partial_x^3}\partial_x^{-1}\Big[\Big(\mathrm{e}^{-t_n\partial_x^3}v(t_n)\Big)\cdot\Big(\partial_x^{-1}\big(\mathrm{e}^{-t_n\partial_x^3}v(t_n)\big)^{k-1}\Big)\Big] + R_2^n.
\end{aligned}
\tag{3.15}
$$

Plugging (3.15) into (3.5), we obtain

$$
\begin{aligned}
&v(t_{n+1})\\
&= v(t_n) - \frac{1}{3}\mathrm{e}^{t_{n+1}\partial_x^3}\partial_x^{-1}\Big[\Big(\mathrm{e}^{-t_{n+1}\partial_x^3}v(t_n)\Big)\cdot\Big(\mathrm{e}^{-\tau\partial_x^3}\partial_x^{-1}\big(\mathrm{e}^{-t_n\partial_x^3}v(t_n)\big)^{k-1}\Big)\Big]\\
&\quad + \frac{1}{3}\mathrm{e}^{t_n\partial_x^3}\partial_x^{-1}\Big[\Big(\mathrm{e}^{-t_n\partial_x^3}v(t_n)\Big)\cdot\Big(\partial_x^{-1}\big(\mathrm{e}^{-t_n\partial_x^3}v(t_n)\big)^{k-1}\Big)\Big] + R_1^n + R_2^n.
\end{aligned}
\tag{3.16}
$$

By inverting the twisted variable $w(t) = e^{-t\partial_x^3}v(t)$ and dropping R_1^n, R_2^n, we find the approximation:

$$w\left(t_{n+1}\right) \approx \Phi\big(w(t_n)\big),$$

with

$$\begin{aligned}\Phi(f) := \mathrm{e}^{-\tau\partial_x^3} f - \frac{1}{3}\partial_x^{-1}\left[\left(\mathrm{e}^{-\tau\partial_x^3} f\right)\cdot\left(\mathrm{e}^{-\tau\partial_x^3}\partial_x^{-1} f^{k-1}\right)\right] \\ + \frac{1}{3}\mathrm{e}^{-\tau\partial_x^3}\partial_x^{-1}\left(f\cdot\partial_x^{-1} f^{k-1}\right).\end{aligned} \tag{3.17}$$

Finally, we invert the gauge transformation. Approximating the time integral in (3.2) by a Riemann sum/left-rectangle rule, we obtain

$$u(t_n, x) \approx w\Big(t_n, x - \frac{\tau}{2\pi}\sum_{j=0}^{n-1}\int_{\mathbb{T}} w(t_j, x)^{k-1} dx\Big).$$

Accordingly, the detailed final form of the scheme is summarized as follows. Denote $u^n = u^n(x) \approx u(t_n, x)$ as the numerical solution of (1.1) for $n \geq 0$. The first-order *gauge-transformed exponential integrator (GTEI)* for solving the gKdV equation (1.1) reads: $w^0(x) = u_0(x)$ and for $n \geq 0$,

$$u^{n+1}(x) = w^{n+1}\Big(x - \frac{\tau}{2\pi}\sum_{j=0}^{n}\int_{\mathbb{T}} (w^j(x))^{k-1} dx\Big), \tag{3.18}$$

$$w^{n+1}(x) = \Phi\big(w^n(x)\big), \quad x \in \mathbb{T}. \tag{3.19}$$

Practically, the above GTEI can be implemented as follows. The propagator Φ in (3.17) in space can be implemented with the help of the FFT, where the computational cost from w^n to w^{n+1} is $\mathcal{O}(N \log N)$ with N the number of spatial grid points. If one is only interested in the value of $u(t, x)$ at a specific time t, GTEI can proceed completely in w^n until $t = t_n$ and then perform the shift in space which can be done by trigonometric interpolations. Such trigonometric interpolations can be accelerated via the non-uniform FFT (NUFFT) [41] if the intermediate values of u within $[0, t]$ are frequently needed. Note that inside the space-shift in (3.18), one only needs the average value of $(w^j(x))^{k-1}$ in space, so there is no need to store the full function $w^j(x)$ for all j. The average of $(w^j(x))^{k-1}$ corresponds to the zero-mode of its Fourier coefficient, or equivalently it can be obtained by a direct Riemann sum on the spatial grids. Since gKdV equations may generate finite-time blowup, e.g., [12, 42] the scheme (3.18) in principle would work before the occurrence of blowup.

3.2. *Main convergence result*

Theorem 1. *Consider the gKdV equation (1.1) for any $k \geq 2$. Let u^n be the numerical solution obtained from the GTEI scheme (3.18) up to some fixed time $T \in (0, T^*)$ with $T^* > 0$ the maximum time for the existence of the solution. Under the assumption that $u_0 \in H^{\gamma+2}(\mathbb{T})$ for some $\gamma \geq 0$, there exist constants $\tau_0, C > 0$ such that for any $0 < \tau \leq \tau_0$ we have*

$$\|u(t_n, \cdot) - u^n\|_{H^\gamma} \leq C\tau, \quad 0 \leq n \leq \frac{T}{\tau},$$

where the constants τ_0 and C depend on T, k and $\|u\|_{L^\infty((0,T);H^{\gamma+2})}$.

The proof is given in [43].

4. Case for classical KdV equation

4.1. *Embedded low-regularity integrator*

We consider the classical KdV equation in this section:

$$\begin{cases} \partial_t u(t,x) + \partial_x^3 u(t,x) + \frac{1}{2}\partial_x\left(u(t,x)^2\right) = 0, \quad x \in \mathbb{T},\ t > 0, \\ u(0,x) = u_0(x), \quad x \in \mathbb{T}, \end{cases} \tag{4.1}$$

for which the regularity requirement can be further reduced thanks to its factorable frequency structure.

4.1.1. *Additional tools*

We denote $\mathbb{P}$ the orthogonal projection onto mean zero functions, that is

$$\mathbb{P}f(x) = f(x) - \frac{1}{2\pi}\int_{\mathbb{T}} f(x)\, dx. \tag{4.2}$$

In particular, we have the special 'formula':

$$\partial_x^{-1}\partial_x f(x) = \mathbb{P}f(x). \tag{4.3}$$

To analyze the resonance of the Fourier frequencies, for simplicity we denote

$$\alpha_{k+1} = \xi^3 - \xi_1^3 - \cdots - \xi_k^3, \quad k \geq 2.$$

Then if $\xi = \xi_1 + \cdots + \xi_k$, we have that

$$\alpha_3 = 3\xi\xi_1\xi_2; \tag{4.4a}$$

$$\alpha_4 = 3(\xi\xi_1\xi_2 + \xi\xi_1\xi_3 + \xi\xi_2\xi_3 - \xi_1\xi_2\xi_3). \tag{4.4b}$$

For convenience of the derivation of schemes, we shall assume that the zero-mode/average of the initial value of (4.1) is zero, that is $\widehat{u_0}(0) = 0$. Otherwise, we may consider instead

$$\tilde{u}(t,x) := u\left(t, x - \widehat{u_0}(0)t\right) - \widehat{u_0}(0), \quad t \geq 0,\ x \in \mathbb{T}, \tag{4.5}$$

and one may note that $\tilde{u}$ also obeys the same KdV equation of (4.1) with initial data $\tilde{u}_0 := u_0 - \widehat{u}_0(0)$. Now by direct calculations, we give the first key fact for our derivation of schemes.

Lemma 2. *(Integration-by-parts)*
(i) Let the space-time functions $f, g \in C^1([0,\tau]; L^2(\mathbb{T}))$ with zero-average $\widehat{f}(t,0) = \widehat{g}(t,0) = 0$ for $t \in [0,\tau]$; then

$$\begin{aligned}
&\int_0^\tau e^{t\partial_x^3}\partial_x\left(e^{-t\partial_x^3}f(t)\cdot e^{-t\partial_x^3}g(t)\right)dt\\
&= \frac{1}{3}e^{\tau\partial_x^3}\left(e^{-\tau\partial_x^3}\partial_x^{-1}f(\tau)\cdot e^{-\tau\partial_x^3}\partial_x^{-1}g(\tau)\right) - \frac{1}{3}\partial_x^{-1}f(0)\cdot\partial_x^{-1}g(0)\\
&\quad - \frac{1}{3}\int_0^\tau e^{t\partial_x^3}\Big[e^{-t\partial_x^3}\partial_x^{-1}\partial_t f(t)\cdot e^{-t\partial_x^3}\partial_x^{-1}g(t)\\
&\qquad\qquad + e^{-t\partial_x^3}\partial_x^{-1}f(t)\cdot e^{-t\partial_x^3}\partial_x^{-1}\partial_t g(t)\Big]\,dt.
\end{aligned}$$

(ii) Let space functions $f_j \in L^2(\mathbb{T})$ (time-independent) be such that $\widehat{f_j}(0) = 0$ for $j = 1,2,3$; then

$$\begin{aligned}
&\int_0^\tau\int_{\mathbb{T}} e^{t\partial_x^3}\left(e^{-t\partial_x^3}f_1\cdot e^{-t\partial_x^3}f_2\cdot e^{-t\partial_x^3}f_3\right)dx\,dt\\
&= -\frac{1}{3}\int_{\mathbb{T}}\left(e^{-\tau\partial_x^3}\partial_x^{-1}f_1\cdot e^{-\tau\partial_x^3}\partial_x^{-1}f_2\cdot e^{-\tau\partial_x^3}\partial_x^{-1}f_3\right)dx\\
&\quad + \frac{1}{3}\int_{\mathbb{T}}\left(\partial_x^{-1}f_1\cdot\partial_x^{-1}f_2\cdot\partial_x^{-1}f_3\right)dx.
\end{aligned}$$

4.1.2. *First-order scheme*

With the above preparation, we now start to derive the first order low-regularity integrator for (4.1). By introducing the twisted variable [34]

$$v(t,x) := e^{t\partial_x^3}u(t,x), \quad t \geq 0,\ x \in \mathbb{T}, \tag{4.6}$$

the KdV equation (4.1) becomes:

$$\partial_t v(t,x) = \frac{1}{2}e^{t\partial_x^3}\partial_x\left(e^{-t\partial_x^3}v(t,x)\right)^2, \quad t \geq 0,\ x \in \mathbb{T}. \tag{4.7}$$

By the Duhamel formula, we have the mild solution for some $n \geq 0$,

$$v(t_n+\tau) = v(t_n) + \frac{1}{2}\int_0^\tau \mathrm{e}^{(t_n+s)\partial_x^3}\partial_x\left(\mathrm{e}^{-(t_n+s)\partial_x^3}v(t_n+s)\right)^2 ds, \quad n \geq 0.$$

Instead of directly approximating $v(t_n+s)$ in the above integrand as was done in the previous work [34], we iterate the Duhamel formula for one time in order to incorporate more inherent structure of the equation. That is, we insert the Duhamel formula of $v(t_n+s)$ into the above equation, and we get

$$\begin{aligned} v(t_{n+1}) = v(t_n) &+ \frac{1}{2}\int_0^\tau \mathrm{e}^{(t_n+s)\partial_x^3}\partial_x\Big[\mathrm{e}^{-(t_n+s)\partial_x^3}\big(v(t_n) \\ &+ \frac{1}{2}\int_0^s \mathrm{e}^{(t_n+\rho)\partial_x^3}\partial_x\left(\mathrm{e}^{-(t_n+\rho)\partial_x^3}v(t_n+\rho)\right)^2 d\rho\big)\Big]^2 ds. \end{aligned} \tag{4.8}$$

We emphasize that this iterative strategy applied here is not aiming for high order accurate approximations as used in the literature (see [34, 37] and some other works in general [25, 26, 36]), but for saving the spatial derivatives, where the nested integral will help to maintain some regularity in the approximation process.

We now begin our numerical approximation based on the above formula. First of all, we let $v(t_n+\rho) \approx v(t_n)$ in (4.8) and we get

$$\begin{aligned} v(t_{n+1}) \approx v(t_n) &+ \frac{1}{2}\int_0^\tau \mathrm{e}^{(t_n+s)\partial_x^3}\partial_x\Big[\mathrm{e}^{-(t_n+s)\partial_x^3}\big(v(t_n) \\ &+ \frac{1}{2}\int_0^s \mathrm{e}^{(t_n+\rho)\partial_x^3}\partial_x\left(\mathrm{e}^{-(t_n+\rho)\partial_x^3}v(t_n)\right)^2 d\rho\big)\Big]^2 ds. \end{aligned} \tag{4.9}$$

By further dropping the $O(\tau^3)$-term in (4.9), we get

$$v(t_{n+1}) \approx v(t_n) + \frac{1}{2}(I_1(t_n) + I_2(t_n)), \quad n \geq 0, \tag{4.10}$$

where we denote

$$\begin{aligned} I_1(t_n) &:= \int_0^\tau \mathrm{e}^{(t_n+s)\partial_x^3}\partial_x\left(\mathrm{e}^{-(t_n+s)\partial_x^3}v(t_n)\right)^2 ds, \\ I_2(t_n) &:= \int_0^\tau\int_0^s \mathrm{e}^{(t_n+s)\partial_x^3}\partial_x\Big[\mathrm{e}^{-(t_n+s)\partial_x^3}v(t_n) \\ &\qquad\qquad \cdot \mathrm{e}^{(\rho-s)\partial_x^3}\partial_x\left(\mathrm{e}^{-(t_n+\rho)\partial_x^3}v(t_n)\right)^2 d\rho\Big]ds. \end{aligned}$$

We will rigorously show later that the dropped $O(\tau^3)$-term in (4.9) can be treated as a $O(\tau^2)$ truncation with only one loss of derivative. As was found in [34] or as a special case of our Lemma 2 (i), the integrand in $I_1(t_n)$ can be exactly evaluated in the physical space:

$$I_1(t_n) = \frac{1}{3}\mathrm{e}^{t_{n+1}\partial_x^3}\left(\mathrm{e}^{-t_{n+1}\partial_x^3}\partial_x^{-1}v(t_n)\right)^2 - \frac{1}{3}\mathrm{e}^{t_n\partial_x^3}\left(\mathrm{e}^{-t_n\partial_x^3}\partial_x^{-1}v(t_n)\right)^2, \ n \geq 0. \tag{4.11}$$

This is very important not only for the accuracy but also for the efficiency of the scheme, since the point-wise formula in the physical space can be efficiently implemented by the Fourier spectral method. However, this is *not* possible for $I_2(t_n)$ unfortunately. Therefore, the following effort is made to approximate $I_2(t_n)$ in order to evaluate it in the physical space, where we pay particular attention to only afford one loss of derivative. The double integral in I_2 does provide flexibility to make this possible as we shall see.

We define for some function w on $\mathbb{T}$ and $s \geq 0$, $n \geq 0$,

$$\begin{aligned} F_n(w,s) :=& \int_0^s \mathrm{e}^{(t_n+\rho)\partial_x^3}\partial_x\left(\mathrm{e}^{-(t_n+\rho)\partial_x^3}w\right)^2 d\rho \\ =& \frac{1}{3}\mathrm{e}^{(t_n+s)\partial_x^3}\left(\mathrm{e}^{-(t_n+s)\partial_x^3}\partial_x^{-1}w\right)^2 - \frac{1}{3}\mathrm{e}^{t_n\partial_x^3}\left(\mathrm{e}^{-t_n\partial_x^3}\partial_x^{-1}w\right)^2, \end{aligned} \tag{4.12}$$

which is obtained similarly as in (4.11), and then $I_2(t_n)$ reads

$$I_2(t_n) = \int_0^\tau \mathrm{e}^{(t_n+s)\partial_x^3}\partial_x\left(\mathrm{e}^{-(t_n+s)\partial_x^3}v(t_n)\cdot \mathrm{e}^{-(t_n+s)\partial_x^3}F_n(v(t_n),s)\right) ds.$$

Noting that $\widehat{F_n(v,s)}(0) = 0$ and applying integration-by-parts in $I_2(t_n)$, with the result of Lemma 2 (i), we find

$$\begin{aligned} I_2(t_n) =& \frac{1}{3}\mathrm{e}^{t_{n+1}\partial_x^3}\left(\mathrm{e}^{-t_{n+1}\partial_x^3}\partial_x^{-1}v(t_n)\cdot \mathrm{e}^{-t_{n+1}\partial_x^3}\partial_x^{-1}F_n\big(v(t_n),\tau\big)\right) & \text{(4.13a)} \\ & - \frac{1}{3}\mathrm{e}^{t_n\partial_x^3}\left(\mathrm{e}^{-t_n\partial_x^3}\partial_x^{-1}v(t_n)\cdot \mathrm{e}^{-t_n\partial_x^3}\partial_x^{-1}F_n\big(v(t_n),0\big)\right) & \text{(4.13b)} \\ & - \frac{1}{3}\int_0^\tau \mathrm{e}^{(t_n+s)\partial_x^3}\Big(\mathrm{e}^{-(t_n+s)\partial_x^3}\partial_x^{-1}v(t_n) \\ & \qquad \cdot \mathrm{e}^{-(t_n+s)\partial_x^3}\partial_x^{-1}\partial_s F_n\big(v(t_n),s\big)\Big) ds. & \text{(4.13c)} \end{aligned}$$

By (4.12), we find

$$\begin{aligned} \text{(4.13a)} =& \frac{1}{9}\mathrm{e}^{t_{n+1}\partial_x^3}\left(\mathrm{e}^{-t_{n+1}\partial_x^3}\partial_x^{-1}v(t_n)\cdot \partial_x^{-1}\big(\mathrm{e}^{-t_{n+1}\partial_x^3}\partial_x^{-1}v(t_n)\big)^2\right) \\ & - \frac{1}{9}\mathrm{e}^{t_{n+1}\partial_x^3}\left(\mathrm{e}^{-t_{n+1}\partial_x^3}\partial_x^{-1}v(t_n)\cdot \mathrm{e}^{-\tau\partial_x^3}\partial_x^{-1}\big(\mathrm{e}^{-t_n\partial_x^3}\partial_x^{-1}v(t_n)\big)^2\right), \end{aligned} \tag{4.14}$$

and since $F_n(v,0)=0$, we have

$$(4.13\text{b}) = 0. \tag{4.15}$$

Now it counts down to handle the term (4.13c) in $I_2(t_n)$. Firstly, note that

$$\partial_s F_n(v,s) = e^{(t_n+s)\partial_x^3}\partial_x\left(e^{-(t_n+s)\partial_x^3}v\right)^2,$$

so by (4.3),

$$\begin{aligned}\partial_x^{-1}\partial_s F_n(v,s) &= \mathbb{P}\left[e^{(t_n+s)\partial_x^3}\left(e^{-(t_n+s)\partial_x^3}v\right)^2\right]\\ &= e^{(t_n+s)\partial_x^3}\left(e^{-(t_n+s)\partial_x^3}v\right)^2 - \frac{1}{2\pi}\int_{\mathbb{T}} e^{(t_n+s)\partial_x^3}\left(e^{-(t_n+s)\partial_x^3}v\right)^2 dx\\ &= e^{(t_n+s)\partial_x^3}\left(e^{-(t_n+s)\partial_x^3}v\right)^2 - \frac{1}{2\pi}\int_{\mathbb{T}} v^2\,dx,\end{aligned}$$

with the operator $\mathbb{P}$ defined in (4.2). Hence, we find

$$\begin{aligned}&(4.13\text{c}) - \frac{\tau}{6\pi}\partial_x^{-1}v(t_n)\int_{\mathbb{T}}(v(t_n))^2\,dx\\ &= -\frac{1}{3}\int_0^\tau e^{(t_n+s)\partial_x^3}\left(e^{-(t_n+s)\partial_x^3}\partial_x^{-1}v(t_n)\cdot\left(e^{-(t_n+s)\partial_x^3}v(t_n)\right)^2\right)ds. \end{aligned}\tag{4.16}$$

Next, we compute the zero Fourier frequency of (4.16) and the others separately. On one hand, by the formula in Lemma 2 (ii), the zero frequency of (4.16) reads

$$\begin{aligned}&\mathcal{F}(4.16)(0)\\ &= -\frac{1}{6\pi}\int_0^\tau\int_{\mathbb{T}} e^{(t_n+s)\partial_x^3}\left(e^{-(t_n+s)\partial_x^3}\partial_x^{-1}v(t_n)\cdot\left(e^{-(t_n+s)\partial_x^3}v(t_n)\right)^2\right)dxds\\ &= \frac{1}{18\pi}\int_{\mathbb{T}} e^{-t_{n+1}\partial_x^3}\partial_x^{-2}v(t_n)\cdot\left(e^{-t_{n+1}\partial_x^3}\partial_x^{-1}v(t_n)\right)^2 dx\\ &\quad - \frac{1}{18\pi}\int_{\mathbb{T}} e^{-t_n\partial_x^3}\partial_x^{-2}v(t_n)\cdot\left(e^{-t_n\partial_x^3}\partial_x^{-1}v(t_n)\right)^2 dx.\end{aligned}$$

By directly checking in the Fourier space, the right-hand side of the above is found simply as

$$\mathcal{F}(4.16)(0) = -\,\mathcal{F}(4.14)(0). \tag{4.17}$$

On the other hand, for $\xi\neq 0$ in the Fourier space of (4.16), we have

$$\begin{aligned}&\mathcal{F}(4.16)(\xi)\\ &= \frac{i}{3}\int_0^\tau\int_{\xi=\xi_1+\xi_2+\xi_3\neq 0}\frac{1}{\xi_1}e^{-i(t_n+s)\alpha_4}\widehat{v}(t_n,\xi_1)\widehat{v}(t_n,\xi_2)\widehat{v}(t_n,\xi_3)\,d(\xi_1)d(\xi_2)ds.\end{aligned}$$

By symmetry, the above is equal to

$$\frac{i}{9}\int_0^\tau\int_{\xi=\xi_1+\xi_2+\xi_3\neq 0}\left(\frac{1}{\xi_1}+\frac{1}{\xi_2}+\frac{1}{\xi_3}\right) \cdot \mathrm{e}^{-i(t_n+s)\alpha_4}\widehat{v}(t_n,\xi_1)\widehat{v}(t_n,\xi_2)\widehat{v}(t_n,\xi_3)\,d(\xi_1)d(\xi_2)ds,$$

and by further noting (4.4b):

$$\begin{aligned}\frac{1}{\xi_1}+\frac{1}{\xi_2}+\frac{1}{\xi_3} &= \frac{1}{\xi_1}+\frac{1}{\xi_2}+\frac{1}{\xi_3}-\frac{1}{\xi}+\frac{1}{\xi}\\ &= \frac{\xi\xi_1\xi_2+\xi\xi_1\xi_3+\xi\xi_2\xi_3-\xi_1\xi_2\xi_3}{\xi\xi_1\xi_2\xi_3}+\frac{1}{\xi}\\ &= \frac{1}{3}\frac{\alpha_4}{\xi\xi_1\xi_2\xi_3}+\frac{1}{\xi},\end{aligned}$$

we find that for $\xi \neq 0$,

$$\begin{aligned}&\mathcal{F}(4.16)(\xi)\\ =&\frac{i}{27}\int_0^\tau\int_{\xi=\xi_1+\xi_2+\xi_3\neq 0}\frac{\alpha_4}{\xi\xi_1\xi_2\xi_3}\mathrm{e}^{-i(t_n+s)\alpha_4}\\ &\qquad\cdot\widehat{v}(t_n,\xi_1)\widehat{v}(t_n,\xi_2)\widehat{v}(t_n,\xi_3)\,d(\xi_1)d(\xi_2)ds \qquad (4.18a)\\ &+\frac{i}{9\xi}\int_0^\tau\int_{\xi=\xi_1+\xi_2+\xi_3\neq 0}\mathrm{e}^{-i(t_n+s)\alpha_4}\widehat{v}(t_n,\xi_1)\widehat{v}(t_n,\xi_2)\widehat{v}(t_n,\xi_3)\,d(\xi_1)d(\xi_2)ds. \qquad (4.18b)\end{aligned}$$

For (4.18a), it is direct to get

$$(4.18a) = -\frac{1}{27}\int_{\xi=\xi_1+\xi_2+\xi_3\neq 0}\frac{1}{\xi\xi_1\xi_2\xi_3}\left(\mathrm{e}^{-it_{n+1}\alpha_4}-\mathrm{e}^{-it_n\alpha_4}\right) \widehat{v}(t_n,\xi_1)\widehat{v}(t_n,\xi_2)\widehat{v}(t_n,\xi_3)\,d(\xi_1)d(\xi_2),$$

which clearly can be transformed back explicitly to the physical space. To further make (4.18b) explicitly integrable in the physical space, we just freeze the phase function in the integrand: $\mathrm{e}^{-i(t_n+s)\alpha_4}\approx \mathrm{e}^{-it_n\alpha_4}$, and simply obtain

$$(4.18b) \approx \frac{i\tau}{9\xi}\int_{\xi=\xi_1+\xi_2+\xi_3\neq 0}\mathrm{e}^{-it_n\alpha_4}\widehat{v}(t_n,\xi_1)\widehat{v}(t_n,\xi_2)\widehat{v}(t_n,\xi_3)\,d(\xi_1)d(\xi_2).$$

This leads to a total $O(\tau^2)$ truncation, but remarkably, thanks to (4.4b) and the factor $\frac{1}{\xi}$ in front, only one spatial derivative is lost in this approximation. Combining the above two findings for (4.18), we obtain that for $\xi \neq 0$,

$$\begin{aligned}\mathcal{F}(4.16)(\xi) \approx& -\frac{1}{27}\int_{\xi=\xi_1+\xi_2+\xi_3\neq 0}\frac{1}{\xi\xi_1\xi_2\xi_3}\left(\mathrm{e}^{-it_{n+1}\alpha_4}-\mathrm{e}^{-it_n\alpha_4}\right)\\ &\qquad\cdot\widehat{v}(t_n,\xi_1)\widehat{v}(t_n,\xi_2)\widehat{v}(t_n,\xi_3)\,d(\xi_1)d(\xi_2)\\ &+\frac{i\tau}{9\xi}\int_{\xi=\xi_1+\xi_2+\xi_3\neq 0}\mathrm{e}^{-it_n\alpha_4}\widehat{v}(t_n,\xi_1)\widehat{v}(t_n,\xi_2)\widehat{v}(t_n,\xi_3)\,d(\xi_1)d(\xi_2).\end{aligned}$$

In total, the inverse Fourier transform of the above together with (4.17) leads to

$$
\begin{aligned}
(4.16) \approx & -\mathcal{F}(4.14)(0) + \frac{1}{27} e^{t_n \partial_x^3} \partial_x^{-1} \left(e^{-t_n \partial_x^3} \partial_x^{-1} v(t_n) \right)^3 \\
& - \frac{1}{27} e^{t_{n+1} \partial_x^3} \partial_x^{-1} \left(e^{-t_{n+1} \partial_x^3} \partial_x^{-1} v(t_n) \right)^3 - \frac{\tau}{9} \partial_x^{-1} e^{t_n \partial_x^3} \left(e^{-t_n \partial_x^3} v(t_n) \right)^3,
\end{aligned}
$$

and this completes our approximation of (4.13c) defined in the physical space.

Finally, combining the above with (4.14) and (4.15) and noting our projection operator $\mathbb{P}$ defined in (4.2), we derive the following approximation from (4.13) as

$$
\begin{aligned}
I_2(t_n) \approx & \frac{1}{9} \mathbb{P} \left[e^{t_{n+1} \partial_x^3} \left(e^{-t_{n+1} \partial_x^3} \partial_x^{-1} v(t_n) \cdot \partial_x^{-1} \left(e^{-t_{n+1} \partial_x^3} \partial_x^{-1} v(t_n) \right)^2 \right) \right] \\
& - \frac{1}{9} \mathbb{P} \left[e^{t_{n+1} \partial_x^3} \left(e^{-t_{n+1} \partial_x^3} \partial_x^{-1} v(t_n) \cdot e^{-\tau \partial_x^3} \partial_x^{-1} \left(e^{-t_n \partial_x^3} \partial_x^{-1} v(t_n) \right)^2 \right) \right] \\
& + \frac{1}{27} e^{t_n \partial_x^3} \partial_x^{-1} \left(e^{-t_n \partial_x^3} \partial_x^{-1} v(t_n) \right)^3 \\
& - \frac{1}{27} e^{t_{n+1} \partial_x^3} \partial_x^{-1} \left(e^{-t_{n+1} \partial_x^3} \partial_x^{-1} v(t_n) \right)^3 \\
& - \frac{\tau}{9} \partial_x^{-1} e^{t_n \partial_x^3} \left(e^{-t_n \partial_x^3} v(t_n) \right)^3 + \frac{\tau}{6\pi} \partial_x^{-1} v(t_n) \int_{\mathbb{T}} (v(t_n))^2 \, dx, \ n \geq 0.
\end{aligned}
$$

This together with the equality for $I_1(t_n)$ in (4.11) ends up as a first order numerical scheme for the mild solution $v^n = v^n(x) \approx v(t, x)$ of (4.7): $n \geq 0$,

$$
\begin{aligned}
v^{n+1} = & v^n - \frac{1}{6} e^{t_n \partial_x^3} \left(e^{-t_n \partial_x^3} \partial_x^{-1} v^n \right)^2 + \frac{1}{6} e^{t_{n+1} \partial_x^3} \left(e^{-t_{n+1} \partial_x^3} \partial_x^{-1} v^n \right)^2 \\
& + \frac{1}{18} \mathbb{P} \left[e^{t_{n+1} \partial_x^3} \left(e^{-t_{n+1} \partial_x^3} \partial_x^{-1} v^n \cdot \partial_x^{-1} \left(e^{-t_{n+1} \partial_x^3} \partial_x^{-1} v^n \right)^2 \right) \right] \\
& - \frac{1}{18} \mathbb{P} \left[e^{t_{n+1} \partial_x^3} \left(e^{-t_{n+1} \partial_x^3} \partial_x^{-1} v^n \cdot e^{-\tau \partial_x^3} \partial_x^{-1} \left(e^{-t_n \partial_x^3} \partial_x^{-1} v^n \right)^2 \right) \right] \\
& + \frac{1}{54} e^{t_n \partial_x^3} \partial_x^{-1} \left(e^{-t_n \partial_x^3} \partial_x^{-1} v^n \right)^3 - \frac{1}{54} e^{t_{n+1} \partial_x^3} \partial_x^{-1} \left(e^{-t_{n+1} \partial_x^3} \partial_x^{-1} v^n \right)^3 \\
& + \frac{\tau}{12\pi} \partial_x^{-1} v^n \int_{\mathbb{T}} (v^n)^2 \, dx - \frac{\tau}{18} \partial_x^{-1} e^{t_n \partial_x^3} \left(e^{-t_n \partial_x^3} v^n \right)^3. \qquad (4.19)
\end{aligned}
$$

By reversing the change of unknown (4.6) in (4.19), we write down the scheme of the *first order embedded low-regularity integrator (ELRI)* for solving the KdV equation (4.1): denoting $u^n = u^n(x) \approx u(t_n, x)$ as the numerical solution, for $n = 0, 1, \ldots$,

$$\begin{aligned} u^{n+1} &= \mathrm{e}^{-\tau\partial_x^3} u^n + A(\tau, u^n) - A(0, u^n) + \frac{\tau}{12\pi} \mathrm{e}^{-\tau\partial_x^3} \partial_x^{-1} u^n \int_{\mathbb{T}} (u^n)^2 \, dx \\ &\quad - \frac{\tau}{18} \partial_x^{-1} \mathrm{e}^{-\tau\partial_x^3} (u^n)^3 \\ &=: ELRI1(u^n), \end{aligned} \tag{4.20}$$

with $\mathbb{P}$ in (4.2), and

$$\begin{aligned} A(s, u^n) &:= -\frac{1}{6} \mathrm{e}^{-s\partial_x^3} \left(\mathrm{e}^{-(\tau-s)\partial_x^3} \partial_x^{-1} u^n \right)^2 \\ &\quad + \frac{1}{18} \mathbb{P} \left[\mathrm{e}^{-\tau\partial_x^3} \partial_x^{-1} u^n \cdot \mathrm{e}^{-(\tau-s)\partial_x^3} \partial_x^{-1} \left(\mathrm{e}^{-s\partial_x^3} \partial_x^{-1} u^n \right)^2 \right] \\ &\quad - \frac{1}{54} \mathrm{e}^{-(\tau-s)\partial_x^3} \partial_x^{-1} \left(\mathrm{e}^{-s\partial_x^3} \partial_x^{-1} u^n \right)^3. \end{aligned}$$

The above ELRI scheme (4.20) is clearly fully explicit and efficient, since it is defined point-wisely in the physical space. All the involved pseudo-differentiation operators and the projection $\mathbb{P}$ in the scheme can be implemented by FFT, where the computational cost per time step is $O(N \log N)$. Also, it is clear that the ELRI (4.20) preserves the mean-value of the solution of the KdV equation (4.1) at the discrete level, i.e.,

$$\int_{\mathbb{T}} u^n(x) dx \equiv \int_{\mathbb{T}} u_0(x) dx, \quad n = 0, 1, \ldots.$$

4.2. *Main convergence result*

Now, we state the convergence theorems of the proposed ELRI1 method (4.20) as the **main results**.

Theorem 3. *Let u^n be the numerical solution from the ELRI1 scheme (4.20) for the KdV equation (4.1) with $\widehat{u_0}(0) = 0$ up to some fixed time $T > 0$. Under the assumption that $u_0 \in H^{\gamma+1}(\mathbb{T})$ for some $\gamma > \frac{1}{2}$, there exist constants $\tau_0, C > 0$ depending only on T and $\|u\|_{L^\infty((0,T);H^{\gamma+1})}$, such that for any $0 < \tau \leq \tau_0$ we have*

$$\|u(t_n, \cdot) - u^n\|_{H^\gamma} \leq C\tau, \quad 0 \leq n \leq \frac{T}{\tau}. \tag{4.21}$$

The proof is given in [38]. A direct result of the above convergence theorems with our shifting technique (4.5) gives the convergence of ELRI for the case of u_0 with general average value in (4.1), i.e., $\widehat{u_0}(0) \in \mathbb{R}$.

Corollary 4. *Let u^n be defined in ELRI1 (4.20) up to some fixed time $T > 0$. Assuming in (4.1) that $u_0 \in H^{\gamma+1}(\mathbb{T})$ for some $\gamma > \frac{1}{2}$, there exist* *constants* $\tau_0, C > 0$ *depending only on* T *and* $\|u\|_{L^\infty((0,T);H^{\gamma+1})}$*, such that for any* $0 < \tau \le \tau_0$,

$$\|u(t_n,\cdot) - u^n\left(\cdot + t_n\widehat{u_0}(0)\right) + \widehat{u_0}(0)\|_{H^\gamma} \le C\tau, \quad 0 \le n \le \frac{T}{\tau}.$$

5. Case for mKdV equation

In this case, when the coefficient in front of the nonlinearity is positive, it is referred to as defocusing, and when it is negative, it is referred to as focusing. In this section, we further consider the case for defocusing mKdV equation:

$$\begin{cases} \partial_t u + \partial_x^3 u = 2\partial_x(u^3), & t > 0,\ x \in \mathbb{T}, \\ u(0,x) = u_0(x), & x \in \mathbb{T}. \end{cases} \tag{5.1}$$

We shall firstly present the connection between the mKdV and KdV equations through the Miura transform, and then based on which we shall derive our numerical approximation. The convergence theorem will be given in the end.

5.1. *Connection with KdV equation*

Let $u = u(t,x)$ be the solution of the mKdV equation (5.1), then by the Miura transformation [44]

$$U := M[u] = \partial_x u + u^2, \tag{5.2}$$

the function $U = U(t,x)$ solves the following KdV equation:

$$\begin{cases} \partial_t U + \partial_x^3 U = 3\partial_x\left(U^2\right), & t > 0,\ x \in \mathbb{T}, \\ U(0,x) = \partial_x u_0 + u_0^2, & x \in \mathbb{T}, \end{cases} \tag{5.3}$$

with u_0 the initial data of the mKdV equation (5.1). By inverting the Miura transform (5.2), we find

$$u(t,x) = \partial_x^{-1}U(t,x) - \partial_x^{-1}u^2(t,x) + \frac{1}{2\pi}\int_{\mathbb{T}} u(t,x)dx, \quad t \ge 0,\ x \in \mathbb{T}.$$

Note that the mKdV equation (5.1) preserves the mass:

$$\int_{\mathbb{T}} u(t,x)dx \equiv \int_{\mathbb{T}} u_0(x)dx =: 2\pi m_0, \quad t \ge 0,$$

so we further get

$$u(t,x) = \partial_x^{-1}U(t,x) - \partial_x^{-1}u^2(t,x) + m_0, \quad t \ge 0,\ x \in \mathbb{T}. \tag{5.4}$$

Thanks to this connection, we are able to call the embedded low-regularity integrator from the previous section for solving the KdV equation (5.3) to get $U(t,x)$, and meanwhile we can make some proper approximation for $u(t,x)$.

To the aim of saving regularity, it is more convenient to work on the twisted variables

$$V = V(t,x) := \mathrm{e}^{t\partial_x^3} U(t,x) \quad v = v(t,x) := \mathrm{e}^{t\partial_x^3} u(t,x), \quad t \geq 0, \ x \in \mathbb{T}, \tag{5.5}$$

respectively for the KdV equation (5.3) and the mKdV equation (5.1). Then, (5.4) becomes

$$v(t,x) = \partial_x^{-1} V(t,x) - \partial_x^{-1} \mathrm{e}^{t\partial_x^3} \left(\mathrm{e}^{-t\partial_x^3} v(t,x)\right)^2 + m_0, \quad t \geq 0, \ x \in \mathbb{T}, \tag{5.6}$$

which is the key relation for us to design numerical scheme.

5.2. *Numerical approximation*

The framework of our approximation for v consists of three steps. We shall first present a coarse approximation for v where some regularity is lost, and then we perform two smoothing iterations based on (5.6) to get some regularity back.

Step 1. A coarse version. We begin by presenting a first coarse approximation for the solution of the mKdV equation (5.1), which can be regarded as a prediction step.

In the spirit of exponential integrators [35], we apply the Duhamel formula to (5.1) in terms of the twisted variable v:

$$v(t_{n+1}) = v(t_n) + 2\int_0^\tau \mathrm{e}^{(t_n+s)\partial_x^3} \partial_x \left[\mathrm{e}^{-(t_n+s)\partial_x^3} v(t_n+s)\right]^3 ds, \quad n \geq 0. \tag{5.7}$$

It can be seen that here in Fourier space, the resonance structure of the phase function for the mKdV equation is

$$\xi^3 - \xi_1^3 - \xi_2^3 - \xi_3^3,$$

which involves the interaction of three frequencies rather than two for the KdV equation. Due to this, the factoring technique used for the KdV equation [34, 37] can not be applied here. Consequently in (5.7), if one only takes the approximation

$$v(t_n+s) \approx v(t_n), \quad 0 \leq s \leq \tau,$$

the rest of the integral cannot be evaluated explicitly in the physical space, and then the triple summation left in the Fourier space would be very costly for computations. Thus, some approximation has to be further made for the integral in (5.7), and unavoidably some more regularity will be lost. For example, one may consider to take

$$\int_0^\tau \mathrm{e}^{(t_n+s)\partial_x^3}\partial_x\left[\mathrm{e}^{-(t_n+s)\partial_x^3}v(t_n+s)\right]^3 ds$$
$$\approx \int_0^\tau \mathrm{e}^{(t_n+s)\partial_x^3}\partial_x\left[\mathrm{e}^{-t_n\partial_x^3}v(t_n)\right]^3 ds, \tag{5.8}$$

and then evaluate the rest exactly. While, this apparently would cost **four** spatial derivatives. What is worse is that such simple approximation will also cause stability issue. Here, we make an effort to save as much regularity as we can, and meanwhile pursue a stable approximation. We replace the Airy operator in (5.8) with a Schrödinger operator, i.e., $\mathrm{e}^{s\partial_x^3}\approx \mathrm{e}^{is\partial_x^2}$. Then, we adopt the following approximation of (5.7) as our first approximation for the mKdV equation:

$$\begin{aligned} v(t_{n+1}) &\approx v(t_n) + 2\mathrm{Re}\int_0^\tau \mathrm{e}^{t_n\partial_x^3+is\partial_x^2}\partial_x\left[\mathrm{e}^{-t_n\partial_x^3}v(t_n)\right]^3 ds \\ &= v(t_n) - 2\mathrm{Re}\left[i\mathrm{e}^{t_n\partial_x^3+i\tau\partial_x^2}\partial_x^{-1}\left(\mathrm{e}^{-t_n\partial_x^3}v(t_n)\right)^3\right] \\ &=: \Phi_I^n(v(t_n)), \quad n\geq 0. \end{aligned} \tag{5.9}$$

This approximation here shares the same spirit as the work in the literature [45–48] to approximate the KdV equation by the Schrödinger equation.

In fact, by denoting the numerical solution $v_I^n\approx v(t_n)$ for $n\geq 0$, (5.9) defines the coarse version of our algorithm for solving the mKdV equation (5.1):

$$v_I^{n+1} = \Phi_I^n(v_I^n), \quad \text{with} \quad v_I^0 = u_0. \tag{5.10}$$

As we shall show rigorously in Section 5.4 that, up to any $T>0$, for $u_0\in H^{\gamma+3}$ with $\gamma>-1/2$, there exists a constant $C>0$ such that

$$\|v(t_n)-v_I^n\|_{H^\gamma}\leq C\tau, \quad 0\leq n\leq \frac{T}{\tau}.$$

That is to say, the first order accurate approximation (5.9) costs **three** spatial derivatives of the solution of the mKdV equation (5.1).

Step 2. Iterative regularizing: first layer. Now with the coarse approximation from the previous step, we introduce an iterative regularizing strategy based on the inverse Miura transform (5.6) to gain the regularity back. This step can be regarded as the first layer in our embedded scheme.

Firstly, the inverse Miura transform (5.6) involves the solution of the KdV equation (5.3). To get this part, we apply the first order embedded low-regularity integrator. Denoting the numerical solution of the KdV equation (5.3) as $U^n \approx U(t_n)$ for $n \geq 0$ with $U^0 = U(0)$, then

$$U^n = \mathrm{e}^{-it_n\partial_x^3}V^n \text{ with } V^n(x) = \widetilde{V}^n\left(x + \frac{3t_n}{\pi}\int_{\mathbb{T}} U^0 dx\right) + \frac{1}{2\pi}\int_{\mathbb{T}} U^0 dx, \tag{5.11}$$

where $\widetilde{V}^0 = \mathbb{P}U(0)$,

$$\widetilde{V}^{n+1} = \widetilde{V}^n + A_n\left(t_{n+1}, \widetilde{V}^n\right) - A_n\left(t_n, \widetilde{V}^n\right) - 2\tau\partial_x^{-1}\mathrm{e}^{t_n\partial_x^3}\left(\mathrm{e}^{-t_n\partial_x^3}\widetilde{V}^n\right)^3$$
$$+ \frac{3\tau}{\pi}\partial_x^{-1}\widetilde{V}^n\int_{\mathbb{T}}\left(\widetilde{V}^n\right)^2 dx,$$

and

$$\begin{aligned} A_n(s,V) = {} & \mathrm{e}^{s\partial_x^3}\left(\mathrm{e}^{-s\partial_x^3}\partial_x^{-1}V\right)^2 - \frac{2}{3}\mathrm{e}^{s\partial_x^3}\partial_x^{-1}\left(\mathrm{e}^{-s\partial_x^3}\partial_x^{-1}V\right)^3 \\ & - 2\mathbb{P}\Big[\mathrm{e}^{t_{n+1}\partial_x^3}\Big(\mathrm{e}^{-t_{n+1}\partial_x^3}\partial_x^{-1}V \\ & \cdot \mathrm{e}^{-(s-t_n)\partial_x^3}\partial_x^{-1}\left(\mathrm{e}^{(s-t_n)\partial_x^3}\mathrm{e}^{-t_{n+1}\partial_x^3}\partial_x^{-1}V\right)^2\Big)\Big]. \end{aligned}$$

Then, by plugging the numerical solution V^n from (5.11) and the prediction algorithm Φ_I^n defined in (5.9) into the right-hand-side of (5.6), we define an improved approximation for the mKdV equation (5.1):

$$v_{II}^n = \partial_x^{-1}V^n - \mathrm{e}^{t_n\partial_x^3}\partial_x^{-1}\left(\mathrm{e}^{-t_n\partial_x^3}v_I^n\right)^2 + m_0, \quad n \geq 1, \quad v_{II}^0 = u_0. \tag{5.12}$$

By doing this, we are able to save one spatial derivative back compared with the coarse version obtained in Step 1. This will be proved rigorously in Section 5.5. With this version, we only need **two** additional spatial derivatives for the first order accuracy, while there is still a room for improvement.

Step 3. Iterative regularizing: second layer. With one more time of iteration where we now embed v_{II}^n from the previous step (5.12) into the right-hand-side of the inverse Miura transform (5.6), we are able to get one more derivative back. That is to say, we now define our final approximation as:

$$v^n = \partial_x^{-1}V^n - \partial_x^{-1}\mathrm{e}^{t_n\partial_x^3}\left(\mathrm{e}^{-t_n\partial_x^3}v_{II}^n\right)^2 + m_0, \quad n \geq 1, \quad v^0 = u_0. \tag{5.13}$$

It can be regarded as the second layer in our embedded scheme. We remark that more iterations will not save any more spatial derivatives, because after the second layer, the regularity requirement of the KdV part will become essential in (5.6). Thus, (5.13) completes our algorithm for solving the mKdV equation (5.1), and its convergence result will be stated as the main theorem below which tells that only **one** additional spatial derivative is needed for the first order accuracy.

5.3. *Scheme and convergence result*

The detailed *ELRI* for solving the mKdV equation (5.1) reads as follows. Denoting $v^n \approx v(t_n)$ as the numerical solution for the twisted variable of the mKdV equation (5.1) for $n \geq 0$ and taking $v^0 = u_0$, then for $n \geq 1$,

$$\begin{aligned} v^n &= \partial_x^{-1} V^n - \mathrm{e}^{t_n\partial_x^3}\partial_x^{-1}\left(\mathrm{e}^{-t_n\partial_x^3} v_{II}^n\right)^2 + m_0, \\ v_{II}^n &= \partial_x^{-1} V^n - \mathrm{e}^{t_n\partial_x^3}\partial_x^{-1}\left(\mathrm{e}^{-t_n\partial_x^3} v_I^n\right)^2 + m_0, \end{aligned} \tag{5.14}$$

where V^n is defined in (5.11) and

$$v_I^n = v_I^{n-1} - 2\mathrm{Re}\left[i\mathrm{e}^{t_n\partial_x^3 + i\tau\partial_x^2}\partial_x^{-1}\left(\mathrm{e}^{-t_n\partial_x^3} v_I^{n-1}\right)^3\right], \quad n \geq 1, \quad v_I^0 = u_0.$$

By inverting the twisted variable, it is equivalent to have the ELRI scheme for the original variable $u(t)$ of the mKdV equation (5.1): denoting $u^n \approx u(t_n)$ for $n \geq 0$ with $u^0 = u_0$, then for $n \geq 1$,

$$u^n = \partial_x^{-1} U^n - \partial_x^{-1}\left(u_{II}^n\right)^2 + m_0, \quad u_{II}^n = \partial_x^{-1} U^n - \partial_x^{-1}\left(u_I^n\right)^2 + m_0, \tag{5.15}$$

where U^n is defined in (5.11) and $u_I^n = \mathrm{e}^{-t_n\partial_x^3} v_I^n$.

The proposed scheme of ELRI (5.14) or (5.15) is fully explicit. In practice, it can be implemented efficiently under the Fourier pseudo-spectral method [49, 50] via the fast Fourier transform. The spatial shifting in (5.11) can be done efficiently by the non-uniform fast Fourier transform [41]. In total, the computational cost per time step is $O(N \log N)$ with $N > 0$ the number of grids points in space.

Now we state our **main result** for the convergence of the proposed ELRI scheme.

Theorem 5. *Let u^n be the numerical solution of the mKdV* (5.1) *obtained from the ELRI scheme* (5.15) *up to some fixed time $T > 0$. Under the assumption that $u_0 \in H^{\gamma+1}(\mathbb{T})$ with $\gamma > \frac{3}{2}$, there exist constants $\tau_0 > 0$ and $C > 0$ depending only on T and $\|u\|_{L^\infty((0,T);H^{\gamma+1})}$, such that for any $0 < \tau \leq \tau_0$,*

$$\|u(t_n, \cdot) - u^n\|_{H^\gamma} \leq C\tau, \quad 0 \leq n \leq \frac{T}{\tau}.$$

The proof can be found in [39]. If one is only interested in the solution $u(t, x)$ of the mKdV equation at the final time $t = T > 0$, the two stages of regularizing at the intermediate time level can be skipped. That is to compute with u_I^n all the way to $n = T/\tau$, and then obtain u^n as (5.15). This could accelerate the practical computations, and the convergence result in Theorem 5 holds at the final time grid.

5.4. *Exponential scheme: The coarse version*

With the preparations from the previous section, we now build the convergence result for the coarse version scheme (5.9) introduced as the first step in Section 5.2. Recall

$$v_I^{n+1} = \Phi_I^n(v_I^n) = v_I^n - 2\mathrm{Re}\left[i\mathrm{e}^{t_n\partial_x^3 + i\tau\partial_x^2}\partial_x^{-1}\left(\mathrm{e}^{-t_n\partial_x^3}v_I^n\right)^3\right],$$
$$v_I^0 = u_0, \quad n = 0, \ldots, \frac{T}{\tau} - 1, \tag{5.16}$$

up to some fixed final time $T > 0$. The convergence result of the coarse scheme (5.16) is stated as follows.

Proposition 6 (Convergence of coarse scheme). *Assume that* $u_0 \in H^{\gamma_1+3}(\mathbb{T})$ *with* $\gamma_1 > -\frac{1}{2}$. *There exist constants* $\tau_0 > 0$ *and* $C > 0$ *depending only on* T *and* $\|v\|_{L^\infty((0,T);H^{\gamma_1+3})}$, *such that for any* $0 < \tau \leq \tau_0$,

$$\|v(t_n) - v_I^n\|_{H^{\gamma_1}} \leq C\tau, \quad 0 \leq n \leq \frac{T}{\tau}.$$

This result serves as an intermediate step for proving our main theorem, and meanwhile it could be of independent interest as we explained in Section 5.2. To prove Proposition 6, we first consider the following local error estimate.

5.5. *Exponential scheme with iterative regularizing*

In this section, we analyze the scheme after the first and the second regularizing iterations in Section 5.2. The technique based on the Miura transform turns out to reduce the loss of regularity. Recall our scheme in the second step of Section 5.2:

$$v_{II}^{n+1} = \partial_x^{-1}V^{n+1} - \mathrm{e}^{t_{n+1}\partial_x^3}\partial_x^{-1}\left(\mathrm{e}^{-t_{n+1}\partial_x^3}v_I^{n+1}\right)^2 + m_0, \; n \geq 0, \; v_{II}^0 = u_0.$$

With the analysis of the coarse scheme, we have the following error estimate for this first regularizing.

Proposition 7. *Assume that* $u_0 \in H^{\gamma+1}(\mathbb{T})$ *with* $\gamma > \frac{3}{2}$. *Then, there exist positive constants* τ_0, C, $\widetilde{C}$ *which depend only on* γ, T *and* $\|v\|_{L^\infty((0,T);H^{\gamma+1})}$, *such that for any* $0 < \tau \leq \tau_0$,

$$\|v_{II}^n\|_{H^\gamma} \leq \widetilde{C} \quad \text{and} \quad \|v(t_n) - v_{II}^n\|_{H^{\gamma-1}} \leq C\tau, \quad 0 \leq n \leq \frac{T}{\tau}. \tag{5.17}$$

In the continuation of the iterative regularizing process, recall from Section 5.2 that after the second regularizing, the scheme reads

$$v^{n+1} = \partial_x^{-1}V^{n+1} - \mathrm{e}^{t_{n+1}\partial_x^3}\partial_x^{-1}\left(\mathrm{e}^{-t_{n+1}\partial_x^3}v_{II}^{n+1}\right)^2 + m_0, \; n \geq 0, \; v^0 = u_0.$$

Its convergence is stated as Theorem 5.

Acknowledgments

Y. Wu is supported by NSFC 12171356 and 11771325. X. Zhao is supported by Natural Science Foundation of Hubei Province No. 2019CFA007 and NSFC 11901440.

References

[1] R. Killip and M. Visan, KdV is well-posed in H^{-1}, *Ann. Math.* **190**, 249–305 (2019).

[2] J. Colliander, M. Keel, G. Staffilani, H. Takaoka, and T. Tao, Sharp global well-posedness for KdV and modified KdV on $\mathbb{R}$ and $\mathbb{T}$, *J. Amer. Math. Soc.* **16**, 705–749 (2003).

[3] J. Colliander, M. Keel, G. Staffilani, H. Takaoka, and T. Tao, Multilinear estimates for periodic KdV equations, and applications, *J. Funct. Anal.* **211**, 173–218 (2004).

[4] J. Bao and Y. Wu, Global well-posedness for the periodic generalized Korteweg-de Vries equation, *Indiana Univ. Math. J.* **66**, 1797–1825 (2017).

[5] C. Courtès, F. Lagoutière, and F. Rousset, Error estimates of finite difference schemes for the Korteweg-de Vries equation, *IMA J. Numer. Anal.* **40**, 628–685 (2020).

[6] C. Klein, Fourth order time-stepping for low dispersion Korteweg-de Vries and nonlinear Schrödinger equation, *ETNA.* **29**, 116–135 (2008).

[7] H. Holden, C. Lubich, and N. Risebro, Operator splitting for partial differential equations with burgers nonlinearity, *Math. Comp.* **82**, 173–185 (2012).

[8] H. Holden, K. Karlsen, N. Risebro, and T. Tao, Operator splitting for the KdV equation, *Math. Comp.* **80**, 821–846 (2011).

[9] H. Holden, K. Karlsen, and N. Risebro, Operator splitting methods for generalized Korteweg-de Vries equations, *J. Comput. Phys.* **153**, 203–222 (1999).

[10] F. Tappert, Numerical solutions of the Korteweg-de Vries equation and its generalizations by the split-step Fourier method. in: (a.c. newell, editor) nonlinear wave motion, *Amer. Math. Soc.* **18**, 215–216 (1974).

[11] J. Bona, V. Dougalis, and O. Karakashian, Fully discrete Galerkin methods for the Korteweg-de Vries equation, *Comput. Math. Appl.* **12**, 859–884 (1986).

[12] J. Bona, V. Dougalis, O. Karakashian, and W. McKinney, Conservative, high-order numerical schemes for the generalized Korteweg-de Vries equation, *Phil. Trans. Royal Soc. London A.* **351**, 107–164 (1995).

[13] B. Guo and J. Shen, On spectral approximations using modified legendre rational functions: Application to the Korteweg-de Vries equation on the half line, *Indiana Univ. Math. J.* **50**, 181–204 (2001).

[14] H. Ma and W. Sun, Optimal error estimates of the Legendre-Petrov-Galerkin method for the Korteweg-de Vries equation, *SIAM J. Numer. Anal.* **39**, 1380–1394 (2001).

[15] J. Shen, A new dual-Petrov-Galerkin method for third and higher odd-order differential equations: application to the KdV equation, *SIAM J. Numer. Anal.* **41**, 1595–1619 (2003).
[16] J. Bona, H. Chen, O. Karakashian, and Y. Xing, Conservative, discontinuous Galerkin-methods for the generalized Korteweg-de Vries equation, *Math. Comp.* **82**, 1401–1432 (2013).
[17] H. Liu and J. Yan, A local discontinuous Galerkin method for the Korteweg-de Vries equation with boundary effect, *J. Comput. Phys.* **215**, 197–218 (2006).
[18] J. Yan and C. Shu, A local discontinuous Galerkin method for KdV type equations, *SIAM J. Numer. Anal.* **40**, 769–791 (2002).
[19] A. de Bouard and A. Debussche, Soliton dynamics for the Korteweg-de Vries equation with multiplicative homogeneous noise, *Elect. J. Prob.*. **14**, 1727–1744 (2009).
[20] M. Gubinelli, Rough solutions for the periodic Korteweg-de Vries equation, *Comm. Pure Appl. Anal.* **11**, 709–733 (2012).
[21] Y. He and X. Zhao, Numerical integrators for dispersion-managed KdV equation, *Comm. Comput. Phys.* **31**, 1180–1214 (2001).
[22] A. Ostermann and K. Schratz, Low regularity exponential-type integrators for semilinear Schrödinger equations, *Found. Comput. Math.* **18**, 731–755 (2018).
[23] B. Li and Y. Wu, A fully discrete low-regularity integrator for the 1d periodic cubic nonlinear Schrödinger equation, *Numer. Math.* **149**, 151–183 (2021).
[24] Y. Li, Y. Wu, and F. Yao, Convergence of an embedded exponential-type low-regularity integrators for the KdV equation without loss of regularity, *Ann. Appl. Math.* **37**, 1–21 (2021).
[25] K. Schratz, Y. Wang, and X. Zhao, Low-regularity integrators for nonlinear Dirac equations, *Math. Comp.* **90**, 189–214 (2021).
[26] M. Knöller, A. Ostermann, and K. Schratz, A Fourier integrator for the cubic nonlinear Schrödinger equation with rough initial data, *SIAM J. Numer. Anal.* **57**, 1967–1986 (2019).
[27] A. Ostermann, F. Rousset, and K. Schratz, Error estimates of a Fourier integrator for the cubic Schrödinger equation at low regularity, *Found. Comput. Math.* **21**, 725–765 (2021).
[28] A. Ostermann, F. Rousset, and K. Schratz, Fourier integrator for periodic NLS: low regularity estimates via discrete bourgain spaces, *J. Eur. Math. Soc.* **25**, 3913–3952 (2023).
[29] A. Ostermann and C. Su, A lawson-type exponential integrator for the Korteweg-de Vries equation, *IMA J. Numer. Anal.* **40**, 2399–2414 (2020).
[30] F. Rousset and K. Schratz, A general framework of low regularity integrators, *SIAM J. Numer. Anal.* **59**, 1735–1768 (2021).
[31] F. Rousset and K. Schratz, Convergence error estimates at low regularity for time discretizations of KdV, *Pure Appl. Anal.* **4**, 127–152 (2022).
[32] Y. Wu and F. Yao, A first-order Fourier integrator for the nonlinear Schrödinger equation on $\mathbb{T}$ without loss of regularity, *Math. Comp.* **91**, 1213–1235 (2022).

[33] Y. Wang and X. Zhao, A symmetric low-regularity integrator for nonlinear Klein-Gordon equation, *Math. Comp.* **91**, 2215–2245 (2022).
[34] M. Hofmanová and K. Schratz, An exponential-type integrator for the KdV equation, *Numer. Math.* **136**, 1117–1137 (2017).
[35] M. Hochbruck and A. Ostermann, Exponential integrators, *Acta Numer.* **19**, 209–286 (2010).
[36] Y. Bruned and K. Schratz, Resonance based schemes for dispersive equations via decorated trees, *Forum Math.* **10**, E2 (2022).
[37] Y. Wu and X. Zhao, Optimal convergence of a second order low-regularity integrator for the KdV equation, *IMA J. Numer. Anal.* **42**, 3499–3528 (2022).
[38] Y. Wu and X. Zhao, Embedded exponential-type low-regularity integrators for KdV equation under rough data, *BIT Numer. Math.* **62**, 1049–1090 (2022).
[39] C. Ning, Y. Wu, and X. Zhao, An embedded exponential-type low-regularity integrator for mKdV equation, *SIAM J. Numer. Anal.* **60**, 999–1025 (2022).
[40] G. Staffilani, On solutions for periodic generalized KdV equations, *Internat. Math. Res. Notices.* **18**, 899–917 (1997).
[41] L. Greengard and J. Lee, Accelerating the nonuniform fast Fourier transform, *SIAM Rev.* **46**, 443–454 (2004).
[42] Y. Martel, F. Merle, and P. Raphaël, Blow up for the critical generalized Korteweg-de Vries equation. I: Dynamics near the soliton, *Acta Math.* **212**, 59–140 (2014).
[43] B. Li, Y. Wu, and X. Zhao, Gauge-transformed exponential integrator for generalized kdv equations with rough data, *SIAM J. Numer. Anal.* **61**, 1689–1715 (2023).
[44] R. Miura, Korteweg-de Vries equation and generalizatiorls. I: A remarkable explicit nonlinear transformation, *J. Math. Phys.* **9**, 1202–1204 (1968).
[45] J. Boyd and G. Chen, Weakly nonlinear wavepackets in the Korteweg-de Vries equation: The KdV/NLS connection, *Math. Comput. Simulation.* **56**, 317–328 (2001).
[46] M. Christ, J. Colliander, and T. Tao, Asymptotics, frequency modulation, and low-regularity illposedness of canonical defocussing equations, *Amer. J. Math.* **125**, 1235–1293 (2003).
[47] G. Schneider, Approximation of the Korteweg-de Vries equation by the nonlinear Schrödinger equation, *J. Differential Equations.* **147**, 333–354 (1998).
[48] T. Tao, Two remarks on the generalised Korteweg de-Vries equation, *Discrete Contin. Dyn. Syst.* **18**, 1–14 (2007).
[49] J. Shen, T. Tang, and L. Wang, *Spectral Methods: Algorithms, Analysis and Applications.* Springer (2011).
[50] L. Trefethen, *Spectral Methods in MATLAB.* SIAM, Philadelphia (2000).

www.ingramcontent.com/pod-product-compliance
Lightning Source LLC
Chambersburg PA
CBHW062243060925
31711CB00016BA/16
* 9 7 8 9 8 1 9 8 1 6 6 1 3 *